TWELVE–FATALITY HOTEL ARSON
RENO, NEVADA

Reported by: Joseph Ockershausen
Harold Cohen, Ph.D.

This is Report 164 of Investigation and Analysis of Major Fire Incidents and USFA's Technical Report Series Project conducted by Tri-Data, a Division of System Planning Corporation under Contract (GS-10-F0350M/HSFEEM-05-A-0363) to the DHS/USFA, and is available from the USFA Web page at http://www.usfa.dhs.gov

 FEMA

Department of Homeland Security
United States Fire Administration
National Fire Data Center

U.S. Fire Administration
Mission Statement

We provide National leadership to foster a solid foundation for local fire and emergency services for prevention, preparedness and response.

Department of Homeland Security
U.S. Fire Administration
Major Fire Investigation Program

The United States Fire Administration (USFA) develops reports on selected major fires throughout the country. The fires usually involve multiple deaths or a large loss of property. But the primary criterion for deciding to write a report is whether it will result in significant "lessons learned." In some cases, these lessons bring to light new knowledge about fire–the effect of building construction or contents, human behavior in fire, etc. In other cases, the lessons are not new, but are serious enough to highlight once again because of another fire tragedy. In some cases, special reports are developed to discuss events, drills, or new technologies or tactics that are of interest to the fire service.

The reports are sent to fire magazines and are distributed at national and regional fire meetings. The reports are available on request from USFA. Announcements of their availability are published widely in fire journals and newsletters.

This body of work provides detailed information on the nature of the fire problem for policymakers who must decide on allocations of resources between fire and other pressing problems, and within the fire service to improve codes and code enforcement, training, public fire education, building technology, and other related areas.

The USFA, which has no regulatory authority, sends an experienced fire investigator into a community after a major incident only after having conferred with the local fire authorities to ensure that the USFA's assistance and presence would be supportive and would in no way interfere with any review of the incident they themselves are conducting. The intent is not to arrive during the event or even immediately after, but rather after the dust settles, so that a complete and objective review of all the important aspects of the incident can be made. Local authorities review USFA's report while it is in draft form. The USFA investigator or team is available to local authorities should they wish to request technical assistance for their own investigation.

For additional copies of this report write to the USFA, 16825 South Seton Avenue, Emmitsburg, Maryland 21727 or visit our website http://www.usfa.dhs.gov

TABLE OF CONTENTS

OVERVIEW 1

SUMMARY OF KEY ISSUES 4

RENO FIRE DEPARTMENT 6

    Building Description. 6

    Building Construction 9

    Fire Detection and Suppression Systems 14

    Offensive Fire Operations 17

    Defensive Fire Operations 21

    Search and Recovery Operations 26

    Fire Load and Spread. 39

EMERGENCY MEDICAL SERVICES 42

    Reno Fire Department. 42

    Regional Emergency Medical Services Authority 42

    Emergency Medical Services Response to the Mizpah Incident 44

FINDINGS AND RECOMMENDATIONS 47

    Fire Department 47

    Emergency Medical Services 49

Mizpah Hotel Fire
Reno, Nevada

Investigated by: Joseph Ockershausen

Harold Cohen, Ph.D.

Project Director: Hollis Stambaugh

Local Contacts: Battalion Chief Bob Knoll

Fire Investigator Tray Palmer

Reno Fire Department

200 Evans Avenue

Reno, NV 89501

(775) 334-2300

Special Agent Dan Heenan, CFI

Bureau of Alcohol, Tobacco, and Firearms

Las Vegas Field Office

Las Vegas, NV 89101

(702) 387-4624

Special Operations Manager Kevin Romero

Regional Emergency Medical Services

Authority (REMSA)

450 Edison Way

Reno, NV 89502

(775) 858-5700

OVERVIEW

On October 31, 2006, a fire occurred in the four-story Mizpah Hotel located at 214 North Lake Street in downtown Reno, NV. Twelve of the hotel's residents died in the fire, 31 people were injured, 2 critically, and 70 people had to be rescued from the building by the fire department. More than 80 people were displaced by the fire.

The fire was detected at approximately 10 p.m., when a smoke detector in the second floor hallway of the hotel's north wing alerted and activated the building's fire alarm system. The hotel fire detection system was a supervised system of smoke and heat detectors monitored 24 hours a day by ADT Security Services in Rochester, NY. The service immediately notified the Reno Fire Department (RFD).[1]

RFD's Station 1 was located approximately 500 feet from the rear of the Mizpah Hotel. The engine and ladder truck from Station 1 were the initial units dispatched to investigate the alarm.[2] As the crews prepared to respond, heavy black smoke was seen coming from the north side of the structure. Prior to leaving the station, the fire officer appraised dispatch of the conditions and requested the incident be upgraded to a working fire assignment for additional assistance.

The first fire units arrived at the scene less than 2 minutes after dispatch. These units were confronted with moderate to heavy black smoke issuing from the windows on the second and third floors of the hotel's north wing. Numerous people were hanging from their windows yelling for help and threatening to jump. The truck captain established Mizpah Command, which was transferred almost immediately to Battalion Chief 1, who arrived immediately behind the first units.

At the time of the fire, the fire alarm system was reported to have functioned properly. The hotel was not equipped with automatic sprinklers, but did have Class I and II standpipe systems. Class I standpipe systems are designed primarily for fire department use, while the Class II system is designed for firefighting by the building occupants until the arrival of the fire department. According to the Incident Commander (IC) one of the hose cabinet lines had been pulled on the second floor.

Many of the hotel residents ignored the fire alarm because alarms had become commonplace in the weeks preceding the fire, and usually were silenced quickly. Many occupants remained in their rooms until they noticed smoke coming under their door, at which time for many it was too late to escape using the interior stairwells or fire escapes. The fire was spreading so rapidly they could no longer leave their rooms safely. Many of those who tried to exit were the first victims. Seven of the 12 victims who perished in the fire were found in the second and third floor hallways.

---

[1] The detection and notification timelines are based on information contained in the fire investigation and fire department incident reports.

[2] This is the fire department's standard response for investigating a fire alarm in the downtown area.

Offensive fire operations were delayed to permit the first-arriving units to rescue people who could be seen at their windows. According to the IC approximately 70 people either were rescued or were escorted out of the building by the fire department; 31 people were reported rescued from upper floor windows by the fire department using ground and aerial ladders. At the height of rescue operations, the fire department used two bucket trucks (also referred to as cherry pickers) provided by the "Contractor's Auxiliary," an organization comprised of local contractors who have agreed to accept the Federal Emergency Management Agency's (FEMA's) fee schedule. The Auxiliary provides a wide variety of heavy and specialized equipment to local jurisdictions during emergencies.

The bucket trucks were used to rescue hotel occupants from their windows in the rear of the building. The decision by the IC to delay offensive fire operations and immediately initiate rescue operations helped to avert additional fatalities and injuries of hotel occupants.

Offensive fire operations lasted for approximately 5 to 10 minutes. Based on the rapid spread and lack of progress at containing the fire, the IC became increasingly concerned as to the structural integrity of the building. He ordered fire crews to evacuate the building immediately, and shifted tactical operations to a defensive strategy. The goal at this point was to prevent the fire from extending from the north wing into the south wing.

Prior to the fire, hotel maintenance personnel had been in the process of replacing mattresses in guest rooms. The old mattresses as well as some new ones were stacked along the walls of the second and third floor hallways waiting to be switched out or disposed of. According to one hotel resident, the mattresses had been positioned along the hallways for about a week. The mattresses significantly contributed to the fire load and spread. According to information on the mattress tags, the padding was 100-percent flexible polyurethane foam, a highly combustible material that can be ignited easily by an open flame such as a lighter or match.

The fire spread rapidly throughout the second and third floors, fed primarily by the polyurethane mattresses stacked along the hallways. The fire extended to the third floor via the stairwells at the west and east ends of the north wing. Virtually all of the rooms on the second and third floors of the north wing were either destroyed or heavily damaged by fire and smoke. Although the annex and south wing suffered little direct fire damage, these areas sustained moderate to heavy smoke damage. The majority of the roof over the north wing was consumed by the fire and eventually collapsed, and a segment of the third floor partially collapsed as well.

A Unified Command was established early into the incident. According to the lead investigator with the Bureau of Alcohol, Tobacco, Firearms and Explosives (ATF) this was one of the best-managed, multiagency major incidents he had seen. The following city, county, and Federal agencies were involved in the incident during fire suppression operations:

- RFD and Command Staff;
- Regional Emergency Medical Services Authority (REMSA);
- Reno Police Department command staff and patrol officers for crowd and traffic control;
- Washoe County District Attorney's Office;
- American Red Cross; and
- City of Reno Building Department.

The fire was investigated by the RFD, the Reno Police Department, Washoe County Sheriff's Office Forensic Investigation Unit, ATF Reno Field Office, ATF National Response Team (NRT), Reno Building Department, and members of the RFD and regional Urban Search and Rescue (US&R) teams.

The fire eventually required four alarms to bring it under control. In total, 16 pieces of fire apparatus and 72 personnel were involved in rescue and fire suppression operations. The fire resulted in $2.4 million dollars in damages. The postfire investigation found that the fire had been deliberately set by a resident of the hotel following an argument with another resident. Both individuals lived on the second floor of the hotel, and both survived the fire.

SUMMARY OF KEY ISSUES

ISSUES	COMMENTS
Fire Origin	The fire originated on the second floor of the north wing when old mattresses stacked in the hallway of the hotel were deliberately set on fire by a resident of the hotel following an argument with a neighboring hotel resident.
Fire Spread	Fire spread was aided by the numerous polyurethane foam mattresses that were stacked along the second and third floor hallways. The mattresses contained 100-percent polyurethane foam padding, which is highly combustible. Once ignited, the foam generated rapid flame spread, intense heat, dense black smoke, and toxic gases. The foam also has a low melting point and, once ignited, melted and formed a combustible liquid, which can migrate from its point of origin, spreading the fire to other areas.
Fire Reporting	There was no delay in reporting the fire. The hotel fire detection system was monitored around the clock by ADT Security Services. ADT received the alarm at 9:59 p.m. and notified the RFD at 10 p.m.
Building Fire Detection and Suppression Systems	The hotel's fire detection system worked properly. However, many residents failed to respond to the fire alarm and remained in their rooms until they saw smoke entering their room. The building was not equipped with an automatic sprinkler system. The building was equipped with a Class I and II standpipe system.
Dollar Loss	$2.4 million.
Casualties	Twelve dead and 31 reported injuries (no fire department injuries).
Escape	The residents who left the building when the alarm sounded were able to use the interior stairwells to reach safety. Some used the four exterior fire escapes. Those who ignored the fire alarm either were rescued by emergency service personnel, or perished in the fire as they later attempted to self-evacuate and were overcome by the intense heat and smoke inside the building.

ISSUES	COMMENTS
Multiple Alarms	The fire resulted in four alarms consisting of nine engine companies, four ladder trucks, two rescue squads, and a breathing air unit. Several units self-dispatched on the fire, which can be dangerous and counterproductive to the IC's Incident Action Plan (IAP), and raises accountability and safety issues.
Staffing Levels	Minimum staffing was four personnel per unit for both engine and ladder trucks. Approximately 72 personnel were involved in the initial fire suppression operations.
Interagency Cooperation	Interagency cooperation was reported to have been excellent. The lead investigator for ATF reported that this was an extremely well-run multiagency incident.
Incident Command System	During fire suppression operations, a Unified Command system was established, consisting of the following agencies: RFD, Reno Police Department, Washoe County Sheriff, REMSA, Washoe County District Attorney's Office, Reno Building Department, members of the RFD and neighboring jurisdictions' US&R teams, Red Cross, and representatives from the public utility companies. The Incident Command was expanded during the postfire investigation to include ATF Reno Field Office, ATF Regional NRT, and support personnel.

RENO FIRE DEPARTMENT

The RFD provides primary fire and emergency services for approximately a 650-square-mile area, which encompasses the City of Reno and the nearby community of Truckee Meadows. Additionally, the RFD provides mutual support and emergency services for nearly 6,000 square miles of Northern Washoe County, NV.

The RFD is comprised of the following resources:

- 17 career fire stations;

- 339 career fire suppression and emergency medical services (EMS) personnel, which includes 120 Emergency Medical Technician Intermediate (EMT-I)-trained firefighters;

- 23 fire prevention and investigation personnel;

- 27 support staff mechanics, supply, and clerical support personnel;

- 18 engine companies, of which 12 have intermediate life-support capabilities;

- 4 ladder trucks;

- 11 wildland fire units;

- 3 water tenders;

- 1 light rescue unit;

- 2 heavy rescue squads with technical rescue capabilities;

- 1 Hazmat mobile laboratory and response unit;

- 3 water rescue response units with rescue boats and kayaks;

- water entry team;

- US&R team; and

- Hazmat team.

The department minimum staffing on all fire suppression units is four personnel, which complies with the staffing recommendations outlined in the National Fire Protection Association (NFPA) Standard 1500, *Standard on Fire Department Occupational Safety and Health Program*.

In 2000, the RFD and Truckee Meadows Fire Protection District entered into a formal agreement in which the RFD administers and coordinates the operational activities of the 11 volunteer/auxiliary fire departments in Truckee Meadows Fire Protection District, and the unincorporated areas of Washoe County. The 11 volunteer/auxiliary fire departments consist of the following:

- 12 engine companies;

- 12 wildland fire units; and

- 4 water tenders.

BUILDING DESCRIPTION

The north and south wings of the Mizpah Hotel bordered North Lake Street to the west. The north wing was bordered by a parking lot, and the south wing was bordered by South Second Street. A service entrance between the north and south wings gave the hotel a "U" shape. A narrow alley traversed the rear of the building, and separated the hotel from a fenced parking lot.

During tactical operations, the RFD uses a lettering system to identify the different sides of the building. Side A of a building is normally the address side or main entrance of the building. Working clockwise, each side of the building received a letter designation i.e., B, C, and D. Figure 1 is an overhead view of the Mizpah Hotel, which illustrates the building layout, geographical orientation, and side designations.

Photo, Courtesy of ATF

Figure 1. Overhead View of Mizpah Hotel before the Fire.

The Mizpah Hotel is not a hotel in the traditional sense. Many of the residents had lived at the hotel for more than 20 years. Under the city's building code, the hotel is classified as an

R-2 residential property that contains sleeping units, or more than two dwelling units in which the residents are primarily permanent in nature. This category includes apartment houses, boarding houses (nontransient), hotels (nontransient), and motels (nontransient) to mention a few. Approximately 82 people lived at the hotel at the time of the fire. Most of them were low-income workers and senior citizens on fixed incomes.

The hotel had 126 rooms, the majority of which (116) were designated as occupant rooms. The majority of the north wing first floor and Annex was comprised of residential rooms. Nonresidential areas on the first floor included a large dayroom, laundry room, storage room, and the hotel lobby. The first floor of the south wing consisted entirely of commercial businesses. The second through the fourth floor of the hotel consisted primarily of resident rooms with common bathroom facilities and several utility and storage areas.

The Mizpah Hotel was not a casino, and was small in comparison to the large modern casino hotels in the city. The hotel's design, size, and construction were consistent with other hotels of its era. The building measured 122 feet wide (north to south), 136 feet deep (east to west), and occupied 0.395 acre of land. The hotel had a total square footage of 59,976 square feet, which included the 45,336-square-foot main building and Annex, and 14,640-square-foot unfinished basement located under the north wing.[3]

The north and south wings of the hotel were three stories in height, and the Annex was four stories. The Annex was attached to the east end of the north wing at ground level, and above the first floor only by narrow enclosed staircases, which can be seen in the following photo between the Annex and north wing. The height difference between the north wing and Annex created a slight offset of approximately 3 to 4 feet between floors of the Annex and north wing. Figure 2 shows the offset in height between the Annex and north wing.

Photo, Courtesy of ATF

Figure 2. Offset Between the Annex and North Wing.

---

[3] Information regarding the buildings dimensions and construction type were taken from the fire investigation report prepared by the ATF

BUILDING CONSTRUCTION

The Mizpah Hotel was built in 1922 by the Ward Brothers, a general contracting firm. When it first opened, the hotel was named the Pincolini. The original hotel consisted of what is now referred to as the north wing or main building. Two subsequent additions doubled the size of the hotel. In 1925, the south wing and front façade connecting the north and south wings was added, followed in 1931 with the addition of the Annex at the rear of the north wing. At the time of the fire, the hotel was undergoing interior renovations, which included the installation of new carpet, and the replacement of old furniture and mattresses acquired from the former Reno Hilton Hotel.[4]

When the north and south wings of the hotel were built, the city had not yet adopted a standard building code. It was not until the late 1920s that the City of Reno adopted a uniform building code. It is not known whether the later additions to the hotel met the city's building code requirements at that time, or what those requirements would have been. The city now uses the 2003 edition of the International Building Code (IBC).

Under the city's current building code, the Mizpah Hotel would be classified as having a Type-III noncombustible construction. Buildings of this construction type usually have exterior walls of a noncombustible material. This type of building construction also is referred to as ordinary construction, which is defined by the NFPA as a structure that has exterior walls, (and structural members that are part of the exterior walls) that are fabricated from noncombustible or limited combustible materials, such as brick or other masonry products. The interior structural members, including walls and columns are built entirely or partially of wood. Such building construction has a fire-resistive rating of 2 hours or less.

The exterior walls of the Mizpah Hotel were of solid masonry construction with parapets of various heights above the roofline, depending on the location of the specific section (the roof had a stairstep design). The hotel's interior walls and ceilings were wood stud construction covered with wood lath and plaster. The floors were hardwood covered with carpeting in the residential areas and hallways, with various tiles in other common areas of the hotel.

The hotel's roof was flat, and comprised of layers of tarpaper and roll roofing over wood decking supported by wood joists. The inverted flat roof design consisted of decking raised several feet above the main roof supported by light wooden supports forming a cockloft, which is a void space between the underside of the roof decking and the ceiling below. Flat roofs have a life span of between 20 and 30 years. It is not uncommon to find numerous layers of tarpaper and other roll roofing 3 or more inches thick in older buildings, where the roof has been built up over many years. The yellow box inside Figure 3 (an aerial photograph) shows the various layers of roofing, including roll roofing, roof planking, and roof joists.

---

[4] Reno Gazette-Journal, November 1, 2006.

Photo, Courtesy of ATF

Figure 3. Overhead Photo Showing the Various Layers of Roofing Material.

Postfire investigation photos taken of the hotel show that the building's inverted "flat" roof actually sloped from front to rear for drainage. This style of roof was common on many large-area roof structures built during the 1920s. The space between the underside of the roof and the ceiling of the top floor is referred to as the cockloft.

Just minutes after the first fire units arrived on the scene, the fire vented through the roof, and spread quickly through the cockloft area unabated, a problem typically associated with flat roofs. A fire in the cockloft may give little indication of its severity until the structural integrity of the roof is compromised, resulting in a localized or catastrophic roof collapse. For this reason, it is extremely important that experienced fire crews be sent to the roof to assess the fire situation and the need for ventilation. If ventilation is necessary, all vertical shafts leading to the roof including interior stairways, skylights, and roof vents should be opened first to release the buildup of heat and smoke inside the building.[5] Although these types of roofs are relatively safe to work on during a fire, firefighters should exercise extreme caution when working on a roof over a fire, and never work on a roof structure that is actually burning. Ventilating a roof increases the airflow to the fire, and a roof structure not involved with fire before ventilation may become involved quickly as soon as the roof is opened.[6] To our knowledge, no attempt was made by the fire department to ventilate the roof, because the majority of personnel were committed to rescuing the trapped occupants. The fire reportedly had vented through the roof during the early stages of the rescue operations, which suggests that the roof may have been weakened by the fire.

---

[5] Fire Engineering. *Fire Officer's Handbook of Tactics*, 3rd Edition.

[6] Brannigan's *Building Construction for the Fire Service*, 4th Edition.

The building was equipped with two skylights. The first was located at the west end of the north wing above the western staircase. This skylight burned away early into the fire as heat and fire advanced up the western stairwell. After the skylight failed, it released an enormous amount of heat and smoke, improving conditions for fire crews advancing up the stairwell to attack the fire. The venting also helped to isolate and stop the fire from spreading into the south wing. The second skylight was located in the roof of the Annex. This skylight appeared to have melted partially, allowing some of the heat and smoke to escape the Annex.

Type III ordinary construction is inherently dangerous, and subject to rapid fire spread and early collapse. More firefighters have been killed in ordinary construction buildings than any other type.[7] Virtually every room on the second and third floors of the north wing was destroyed by the fire or suffered significant smoke damage. The roof over the north wing was totally consumed. Early into offensive fire operations the roof partially collapsed, followed by a partial collapse of the third floor, as well. The fire damage to the north wing was so extensive that fire investigators delayed their postfire investigation of the third floor, until the local US&R team completed shoring operations to stabilize the third floor of the north wing.

The hotel had three interior staircases. The north wing had an interior staircase (west staircase) at the front that accessed the first to the third floor of the hotel. The east staircase at the rear of the north wing (Side C) accessed the first through the fourth floors of the Annex, and the third floor of the north wing.

The second and third floors of the north wing were similar in layout, with two parallel hallways terminating at stairwells at either end. The western staircase was the only stairway in the hotel that led directly to the exterior of the building through the main entrance on Side A of the hotel. A third central staircase was located at midway of the west hallway between the north and south wings. This staircase accessed the residential rooms on the second and third floors of the south wing. A single central hallway provided access to all rooms in the south wing. Figure 4 shows the basic floor plan of the second through fourth floors of the Annex and north and south wings.

---

[7] Brannigan's *Building Construction for the Fire Service*, 4th Edition.

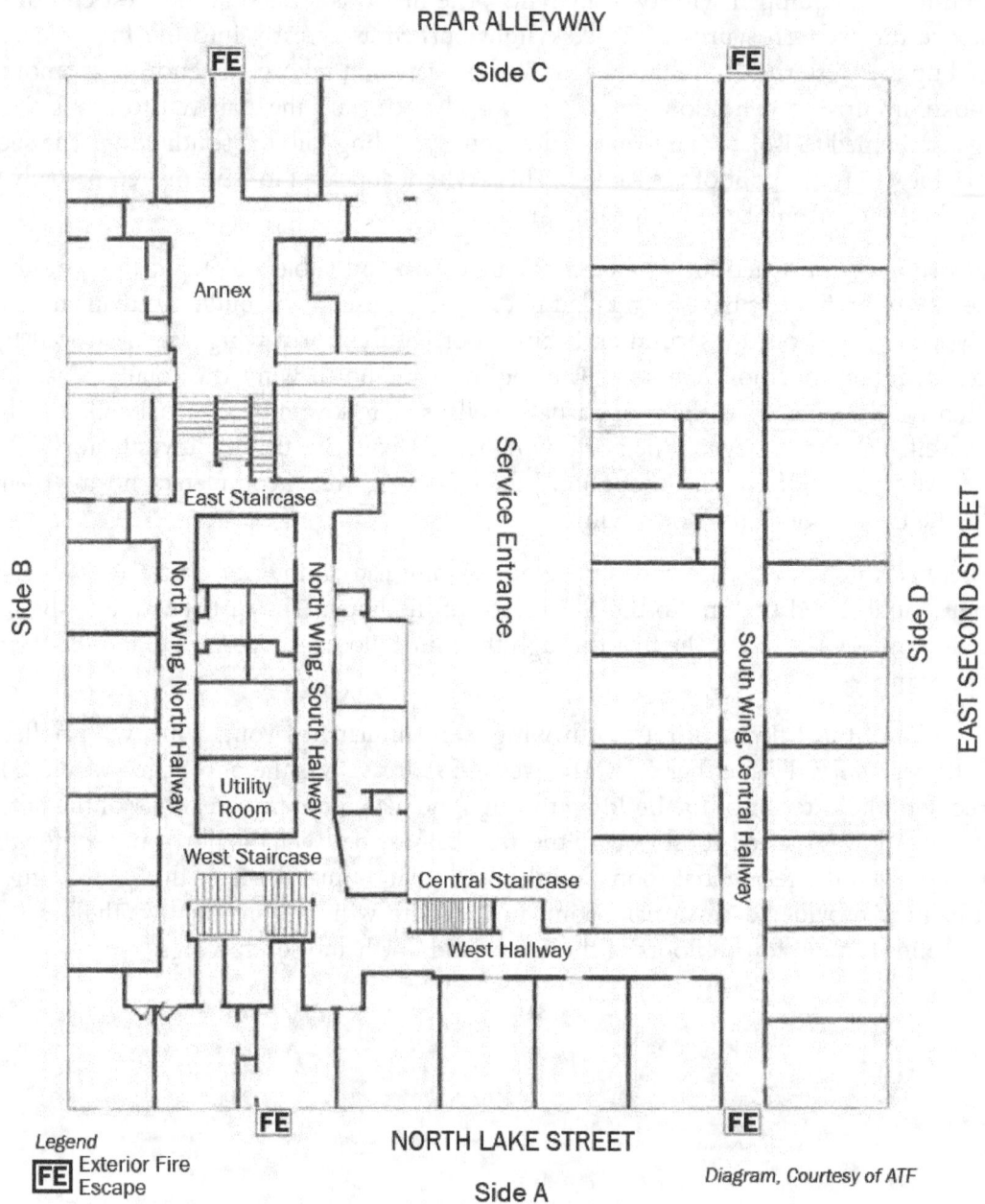

Figure 4. Basic Floor Plan of Annex and Hotel.

The hotel had four exterior fire escapes. Two were located on the front (Side A) of the hotel at the west end of the north and south wings, and two at the rear of the building that serviced the south wing and Annex. The north and south wing fire escapes serviced the second and third floors of the hotel. The fire escape at the rear of the Annex serviced the second through the fourth floors of the Annex. **The only access to the fire escapes was through the interior hallways.** During the fire, the interior conditions deteriorated so rapidly that the interior hallways and stairwells quickly became untenable and residents who failed to exit the building immediately could not get to the fire escapes and to safety. Figure 5 and Figure 6 show the location of the exterior fire escape on the front and rear of the hotel.

Photo, Courtesy of ATF

Figure 5. Front Fire Escapes.

Figure 6. Rear Fire Escapes.

FIRE DETECTION AND SUPPRESSION SYSTEMS

The hotel was equipped with supervised fire detection and alarm systems monitored around the clock by ADT Security Services in Rochester, NY. The systems were divided into eight zones, which covered the entire hotel. Investigators during the postfire investigation noted that the zone coverage areas were too large. Large fire zones can cause fire crews to take a significant amount of time to locate the fire, giving it a chance to spread. Table 1 shows the eight zones and corresponding coverage areas.

Table 1. Alarm Coverage Zones.

Alarm Zones	Coverage Area
Zone 1	Lake and Second Street businesses
Zone 2	Lake and Second Street businesses
Zone 3	Rooms 50 to 57 second floor Annex Rooms 59 to 64 third floor Annex
Zone 4	Rooms 23 to 43 third floor north wing Rooms 65 to 72 fourth floor Annex
Zone 5	Rooms 98 to 117 third floor south wing
Zone 6	Basement and house laundry room
Zone 7	Lobby and first floor halls
Zone 8	Rooms 1 to 19 second floor north wing Rooms 21 to 97 second floor south wing

The alarm systems consisted of 7 manual pull stations, 17 hard-wired photoelectric smoke detectors, and 10 heat detectors. The smoke detectors were mounted at ceiling level near the end and the middle of the hallways. The residential rooms were equipped with battery-operated smoke detectors, which were not interconnected with the building's interior alarm system.[8] The heat detectors were located throughout the building and consisted of a mixture of fixed-temperature (194 °F (90 °C)) and rate-of-rise (135 °F (57 °C) +/- 5/five minutes' detectors). The audible warning devices included six horns and six strobe lights. The manual pull stations were located near the stairwells and in some hallways. (In contrast, new hotels are required to be fully sprinklered, and have smoke detectors in every room.) Despite the detection and alarm system's limitations, the system functioned properly, and ADT quickly notified the fire department.

The system first alerted to the fire at 9:59 p.m. when a smoke detector in Zone 8 activated. This zone covered the entire second floor of both the north and south wings. Fire investigators determined the fire started on the second floor of the north wing. In less than 2 minutes detectors in Zones 3, 4, and 5 activated as well, which is a good indication of how fast the smoke and heat were spreading. Figure 7 shows the second floor area encompassing Zone 8 in which the first smoke detector alerted. Table 2, which follows, shows the time sequence of smoke detector and alarm activations the evening of the fire.

---

[8] All of the information pertaining to the fire detection and alarm systems was taken from the RFD fire investigation report.

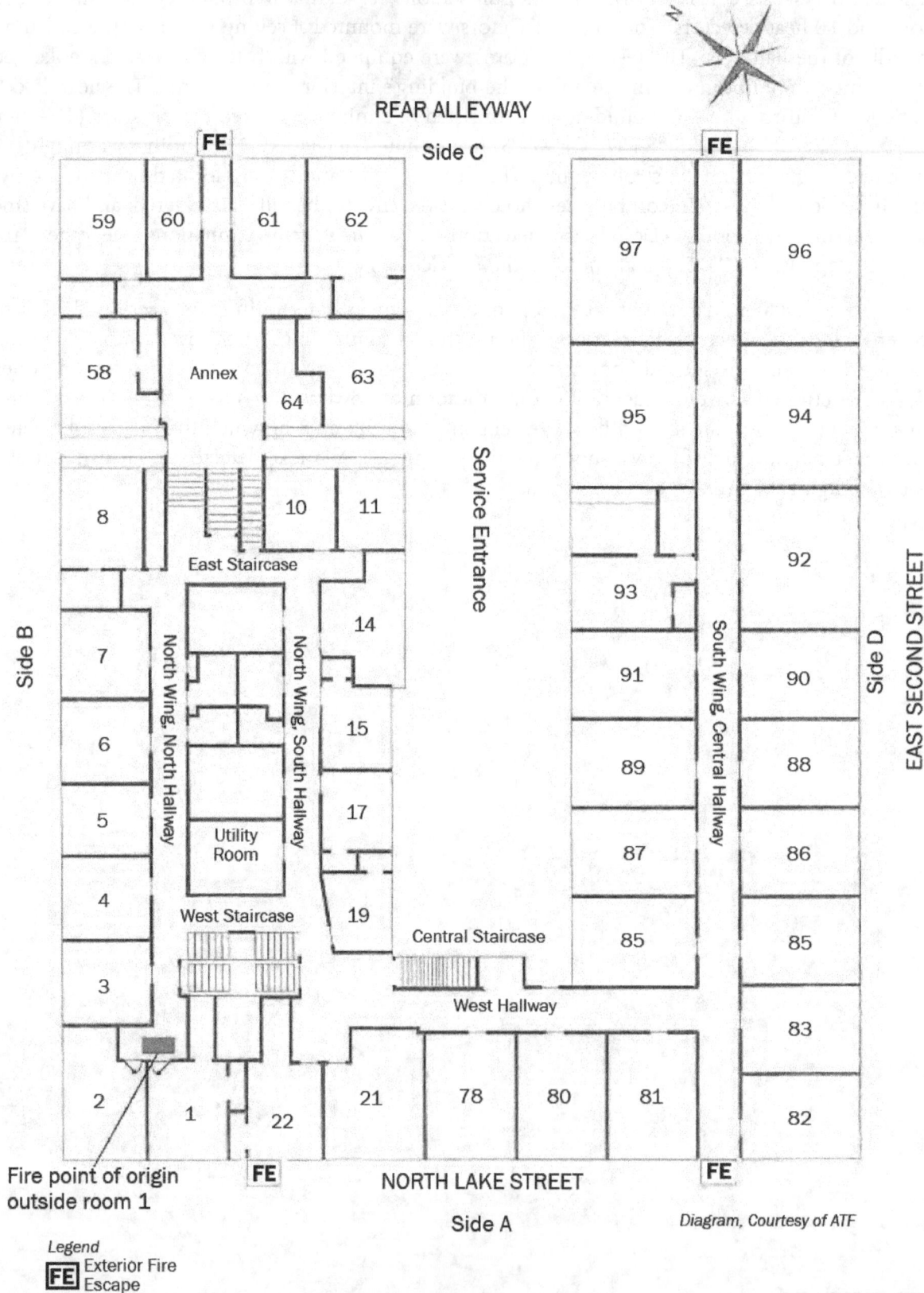

Figure 7. Fire Zone 8.

Table 2. Time Sequence of Alarm Activations.

Alarm Zones	Coverage Area	Location
2158.58	Zone 8	2nd Floor, Rooms 1 to 19 North Wing, Rooms 21 to 97 South Wing
2159.38	**N/A**	**ADT notifies the Reno Fire Department**
2200.35	Zone 3	2nd Floor Annex, Rooms 50 to 57, 3rd Floor Annex, Rooms 58 to 64
2200.43	Zone 4	3rd Floor, Rooms 23 to 43 North Wing, Rooms 65 to 72 4th Floor Annex
2200.47	Zone 5	3rd Floor, Rooms 98 to 117 South Wing
2200.53	Zones 3 and 4	Trouble (indication of compromised circuit)
2201.33	Zone 6	Trouble
2201.33, 37 and 41	Zones 7, 1, and 2	Trouble
2202	N/A	ADT unable to contact premises
2204.26	Still Receiving Signals	Operator activity only

The building was not equipped with sprinklers. When the hotel was built, there was no legal requirement for an automatic sprinkler system. Following adoption of the city's first building code, owners were not obligated to retrofit the hotel with a sprinkler system, since the use group and ownership had not changed. A sprinkler system would be required only if there were to be a change in use group.

The hotel was equipped with Class I and Class II standpipe systems. Class I standpipe systems are designed for fire department use only. According to the fire department's chief fire inspector, the only Class I standpipe system in the hotel was in the west stairwell (Side A) of the north wing. That system fed off the domestic water supply, but was not used during the fire because the western stairwell was compromised by heat and heavy smoke. Class II standpipe systems are designed for occupants' use until the arrival of the fire department. At the Mizpah, that system was last tested in March of 2006 and was reported to have passed inspection. The IC reported that at least one hoseline from a cabinet had been pulled on the third floor, and was found lying in the hallway of the south wing. It was not known who deployed the line or whether the line had been charged.

OFFENSIVE FIRE OPERATIONS

As noted, the fire communication center received the automated alarm at approximately 10 p.m. At 10:02 p.m., fire units from Station 1 (headquarters station) were dispatched to investigate the alarm.

Station 1 is located approximately one block from the hotel, which could be seen from the fire station. The initial fire response consisted of an engine company (**E-1**) with four personnel and a ladder truck (**T-1**) also staffed with four personnel. This is the RFD's standard response to an automatic fire alarm in the city's downtown area.

As fire units prepared to leave the station, the officer (captain) assigned to **T-1** observed heavy black smoke issuing from the north side (Side B) of the hotel. The captain informed dispatch of the heavy smoke conditions, and immediately requested that the initial response be upgraded to a third alarm. Dispatch asked the captain if he would prefer a highrise assignment, and he acknowledged yes.

Prior to the fire department's arrival, the assistant manager of the hotel attempted to extinguish the fire with two portable fire extinguishers, but was unsuccessful.

T-1 and **E-1** arrived on the scene at 10:04 p.m. and the officer aboard **T-1** immediately established Mizpah Command in front of the building.[9] At approximately the same time, fire department communications dispatched the highrise assignment. For the purpose of this report, the highrise assignment is referred to as the second alarm. The second alarm consisted of a Command Officer, **B/C-1**, **RS-1, E-2, E-3, E-5, T-3,** and a Safety Officer.

Although a Command Officer was not dispatched on the initial alarm, **B/C-1** immediately responded after hearing the truck officer's description of the smoke conditions. The **B/C** arrived almost simultaneously with the first-arriving fire units and immediately took Command and established the Incident Command Post (ICP) on North Lake Street opposite Side A of the hotel. He reported a working fire, which initiated the automatic response of additional nonfirefighting resources: law enforcement assistance for crowd and traffic control, air unit to replenish self-contained breathing apparatus (SCBA) equipment, utility company officials, the oncall fire investigator, and a notification of the department's other onduty Command Officers.

The first-arriving units were confronted with heavy black smoke issuing from the third floor windows on Side B of the north wing, and light smoke showing from several windows on the second floor. Numerous people were leaning out of their room windows yelling for help and threatening to jump. Police officers and bystanders were attempting to calm and rescue the people they could.

E-1 took a position on the northwest corner of the north wing leaving room for the aerial ladder in front of the building. Bystanders in a nearby parking lot immediately alerted the officer aboard **E-1** of 8 to 10 people at their windows on the second and third floors. The situation was so critical that the initial response was directed exclusively at rescue operations per standard procedure; offensive operations were delayed to allow fire crews time to assist the police and by-standers with the rescues. Despite the presence of heavy smoke, the first-arriving engine did not lay a supply line.

T-1 positioned in the middle of North Lake Street at the northwest corner of the hotel. Operating from this location the aerial ladder could easily cover large segments of Sides A and B thus maximizing the aerial ladder capabilities.

Using ground ladders, **E-1** began to rescue people from windows on Side B (north Side) of the building. The rescues were prioritized, with the most endangered rescued first. Using the aerial ladder **T-1** rescued an individual from a third-floor window on Side B of the north wing. A female occupant was rescued over a ground ladder from a second-floor window on Side A of the north

---

[9] Much of the response data contained in the incident report appeared to be incorrect. The response and arrival times that appear in the offensive and defensive operation sections of this report are approximate, based on estimated response time for fire apparatus provided by the IC.

wing.[10]   During this rescue, the truck captain observed someone waving a light from inside a third-floor room. After the woman had been safely removed from the building, the crew immediately repositioned the ground ladder and made entry through the third floor window where they located a male victim who was slipping in and out of consciousness. The man was very heavy, and it took both firefighters approximately 10 minutes to position the man at the window's ledge and remove him down the **T-1** aerial ladder.

Personnel from **T-1** also secured the building's utilities. Later, utility crews from Sierra Pacific Power deactivated the overhead power lines that had caused some safety problems for rescue personnel and firefighters operating a master stream in the rear (Side C) of the hotel.

R-1 also responded on the initial alarm. The crew self-dispatched with the engine and truck from Station 1, and arrived on the scene at approximately the same time. The unit was positioned on Side A opposite the northwest corner of the hotel. The two-person crew was directed by the captain on **T-1** to rescue an elderly man from a second floor window of the north wing. Following the rescue, the man was handed over to EMS personnel to evaluate his medical condition. As rescue efforts continued, the crew on **R-1** successfully rescued two additional people, one from a third-floor window on Side B and a second from a window on Side C of the Annex.

While returning to quarters from a medical emergency, **E-4** had been monitoring the radio traffic and also self-dispatched to the incident arriving at 10:05 p.m. Normally, **E-4** would not have responded on the call until a later alarm. The officer aboard **E-4** reported that as they approached the scene, fire was visible from windows on Side B. **E-4** was assigned to establish a water supply to **E-1**, and then initiate an interior attack on the fire.

Conditions inside the building were deteriorating so rapidly that many of the hotel residents were no longer able to access the interior stairwells or fire escapes to exit the building. At the time, the fire department was unaware of the numerous polyurethane foam mattresses positioned along the interior hallways. During an interview, the IC commented that it appeared that a flammable accelerant may have been used because of the thick black smoke and rapid spread of the fire. Many rooms were filling with deadly concentrations of smoke. The doors of the occupant rooms were thin, nonrated wood-panel doors. Some room doors failed, allowing the fire to extend from the hallway into residential rooms while some occupants were waiting to be rescued.

As additional fire units arrived at the fire scene, fire suppression operations were initiated and conducted simultaneously with rescue operations in an attempt to slow progression of the fire and to reduce the danger to those who had not been rescued.

E-4 completed the water supply from the fire hydrant at the corner of East Commercial Row and North Lake Street, 400 feet away to **E-1**. The crew then split up and assisted **T-1** with two rescues; one from a second floor window of the north wing, and another from a second floor window on Side B of the Annex.

---

[10] This individual later admitted to setting the deadly fire.

While the crew from **E-1** was engaged in rescue operations, the driver advanced a highrise pack to the main entrance of the hotel. The pack consisted of a 3-inch truck line connected to a gated wye that supplied two 100-foot, 1-3/4-inch attack lines with fog nozzles. The two handlines had a combined flow capacity of approximately 350 gallons per minute (gpm).

As the second-alarm units began to arrive on the fire scene they were assigned to assist with the ongoing rescue effort. **E-2** arrived on the fire scene at 10:07 p.m. and took a position on the southeast corner of the hotel at the entrance to the alley in the rear of the hotel (Side C). The IC first assigned **E-2** to initiate an interior attack. However, the crew advised Command of the rescue problem that existed in the rear of the building, and the IC reassigned **E-2** to assist in the rescue efforts.

E-2's crew split up. One firefighter assisted personnel from **E-1** in removing a trapped occupant from a third floor window where heavy smoke was issuing. This individual was handed over immediately to EMS personnel (REMSA) at the scene. REMSA personnel initiated life support measures, but the victim failed to respond and was pronounced dead on the scene. This individual was the first reported fatality of the fire. At the same time **E-2's** other crew members removed a ground ladder from their engine and rescued two people from a window on the northeast corner of the hotel. After making the rescue, the two firefighters returned to the third floor room and attempted to conduct a search of the room until they were forced from the room by intense heat and smoke. According to the IC, two double-bucket trucks (cherry pickers) were requested through the local Contractor's Auxiliary. These units were used in lieu of aerial ladders to remove several occupants (number unknown) from Side C of the Annex due to overhead powerlines at the rear of the building. According to the Operations Chief, Auxiliary personnel operated the bucket trucks, and firefighters coached the trapped occupants into the buckets. Although the use of bucket trucks was an innovative approach in overcoming a rescue problem, it raises other safety issues such as the safe-weight operating capacity of the units, and the operational limitations of the vehicles. The double-bucket trucks reportedly had a combined weight limitation of 600 pounds, which is pushing the envelope when a third person is introduced into the equation. During these rescues, one individual jumped from a window in the alcove (service entrance area) into a dumpster in the service area and sustained minor injuries. REMSA evaluated and treated the individual.

After completing the water supply to **E-1**, three personnel from **E-4** advanced one of the 1-3/4-inch handlines from the standpipe pack through the hotel main entrance and checked the first floor where conditions were found to be clear. The crew then advanced the line up the west staircase of the north wing. At the top of the staircase the firefighters encountered heavy smoke and fire conditions so severe that the doors leading from the stairwell to the second floor hallways had been partially burned away. The crew knocked down the fire and moved into the north hallway. The Company Officer (CO) later reported that the fire was in such a free-burning state that visibility was excellent. He could actually see the entire length of the hallway, including the damage the fire had caused to the ceiling and walls of the hallway and doors. The heat was so intense that the firefighters' helmets began to melt.

Firefighters left the west stairwell on the second floor and knocked down the fire at the west end of the hallway. As they moved toward the rear of the building attacking the fire, firefighters encountered mattress springs on the floor of the hallway. These were remnants of the mattresses that the hotel had been in the process of replacing. Approximately a week prior to the fire, hotel maintenance personnel had replaced mattresses in some of the residential rooms. As the old mattresses were removed, they were placed along the walls of the hallways of the second and third floor.

According to hotel maintenance personnel, on the day of the fire 17 mattresses were stacked laterally along the walls of the second-floor hallway of the north wing. These were twin-size mattresses reported to have 100-percent polyurethane foam padding and other types of filling materials.[11] Polyurethane foam can produce rapid flame spread, intense heat, dense black smoke, and toxic gases. All of these burning characteristics were encountered during this fire.

E-3 arrived on the scene at 10:08 p.m. and was assigned by Command to assist E-4 in attacking the fire on the second floor. The crew advanced the second 1-3/4-inch line from the highrise pack to the second floor and entered the west hallway opposite E-4's position. This hallway joined the north and south wings of the hotel on Side A. E-3 encountered heavy fire conditions as they entered the west hallway. The officer later recounted that the main body of fire was advancing from behind at a rapid pace. He described the fire as rolling along the ceiling behind and in front of them in the direction of the center staircase. E-3 quickly knocked down the approaching fire and used their hoseline to ventilate the smoke hydraulically from a second-floor window to improve visibility.

E-5 and T-3 arrived on the fireground at 10:08 p.m. T-3 took a position at the southwest corner of the building. From this position the aerial ladder could reach Sides A and D of the building. The truck was assigned to assist the other crews with the ongoing rescue operations. The crew assisted with other units with ground ladder rescues in the rear of the building, and the aerial ladder was used to rescue two people from a third-floor window of the south wing, and a third person from another window, also on the third floor.

E-5 was assigned to assist E-2 on the second floor in attacking the fire. As E-5 moved into position to back up E-3, Command issued the order for all units to evacuate the building and conduct a personnel accountability report (PAR). E-3 reported that they abandoned their line and left the building. All personnel were accounted for.

Several city buses were secured and used as temporary shelters for individuals displaced by the fire. The local chapter of the American Red Cross (ARC) opened a shelter at the Reno High School gymnasium, where 56 residents from the hotel were taken until alternative housing could be arranged. The Logistics Officer was requested to obtain portable sanitary facilities and food and beverages to rehabilitate emergency personnel.

DEFENSIVE FIRE OPERATIONS

Even after about 10 minutes of offensive fire operations, the fire continued to spread rapidly. Smoke and fire became visible from the second- and third-floor windows of the north wing, and the fire had vented through the roof. Little if any progress was being made in extinguishing the fire. It was reported that Harrah's Hotel and Casino located directly across the street from the Mizpah Hotel considered evacuating the portion of the hotel closest to the fire; however, they did not actually evacuate any guests.

A generally accepted rule in the fire service is that when a sufficient number of hose-streams and adequate ventilation has failed to bring a fire under control within 20 minutes, the IC should con-

---

[11] Information concerning the polyurethane foam padding used in the manufacture of the mattresses involved in the fire was taken from the ATF fire investigation report. The burning characteristics and hazards associated with polyurethane foam was obtained from a Material Safety Data Sheet (MSDS) published by Foamex International Incorporated.

sider evacuating personnel from the building, and switch from an offensive to a defensive mode of operation. In the case of the Mizpah Hotel, the fuel load associated with the numerous polyurethane foam mattresses stacked along the hallways created a more intensely burning and rapidly spreading fire that accelerated the IC's decision to shift to defensive operations earlier.[12]

Since there was so little progress in extinguishing the fire, the IC became concerned about the structural stability of the building. He ordered all emergency personnel out of the building. Communications activated the emergency alert tones and announced for personnel to evacuate the building. Following the evacuation, all companies were ordered to account for their personnel and render a PAR. All personnel were safe.

Tactical operations then shifted from an offensive to a defensive strategy. The primary goal at this point was to prevent the fire from extending into the south wing of the hotel. Units already on the fire scene were ordered to commence master-stream operations. Ladder trucks were rigged for ladder pipe operations, and deluge sets and large-caliber handlines were positioned along the north and west sides of the north wing to attack the fire.

A coordinated attack using exterior handlines and master streams was used. The crews from **E-1, E-4,** and **E-5** deployed four 2-1/2-inch exterior handlines on Side B of the north wing, attacking the fire through the windows on the second and third floors. The crew from **E-3** placed the deck gun on **E-1** in service and later deployed a monitor pipe from **E-3** and began attacking fire from Side A of the building. This helped to keep the fire from spreading through interior hallways into the south wing. **E-1** maintained a defensive position until relieved for rehab at 2:30 a.m. Following rehab, **E-1** remained on standby status until placed in service at 6 a.m.

T-1 was tasked with preventing the fire from spreading to the south wing. Pumped by **E-1, T-1** initiated master-stream operations and attacked the fire through the third-floor windows on Side A, and then directed their ladder pipe to the roof structure to extinguish most of the fire in the roof, thus preventing the fire from penetrating the third-floor firewall separating the north and south wings. **T-1** was successful in knocking down a large volume of the fire. After the bulk of the fire was extinguished, **T-1** continued to monitor the north wing, and extinguished any flareups that occurred. **T-1** also retrieved the hotel registry and file drawers to gain information about the number of residents residing in the hotel at the time of the fire. This information was turned over to the Command Post (CP) in an effort to account for all hotel residents.

Upon concluding rescue operations, **E-2** maintained position in the rear of the building in case anyone else required rescue. Police officers in the rear of the building continued to use their public address systems to direct anyone who still may have been inside the building to go to windows for rescue. No one responded. More than 30 rescues were made by emergency service personnel and utility crews using their bucket trucks. Only one of the more than 30 occupants who were rescued died.

T-3 arrived at the fire scene at 10:11 p.m. According to the officer aboard **T-3**, the fire initially vented through the roof near the center of the north wing and was spreading rapidly throughout the roof. The truck was directed to protect the south side of the firewall separating the north and south wings.

---

[12] Fire Engineering. Fire Officers Handbook of Tactics, 3rd Edition.

A small trench cut was made as an observation port along the south side of the firewall. At 15-minute intervals, a firefighter climbed the ladder and checked to ensure the fire had not penetrated the firewall. Once it was determined that the fire was under control, Command directed all master streams to concentrate on extinguishing the main body of fire.

A third alarm was requested by the IC to assist with defensive operations. The third-alarm units included a second Command Officer, **B/C-2** who was designated the Operations Officer, the Breathing Air Unit, **E-4** (already on the scene), **E-10, E-35, T-11**, three Advance Life Support (ALS) units, an EMS supervisor, and a Safety Officer.

E-10 arrived on the scene at 10:18 p.m. and took up position at the fire hydrant on the corner of Lake and Second Streets. **E-10** pumped **T-3's** master stream. Both crews were later reassigned to assist with search and recovery operations, which will be discussed later in the report.

T-11 arrived on the scene at approximately 10:19 p.m. and was ordered to stage in the parking lot at the rear of the building and stand by for possible master-stream operation. **E-2** repositioned to the fire hydrant at the corner of Evans and Second Streets and supplied **T-11** with water. The truck was supplied through approximately 250 feet of 5-inch large-diameter hose. However, due to overhead power lines that traversed to the rear of the building, **T-11** was unable to flow water until the power company had shut off the power to the building. **T-11** stood by unused until they eventually were released by Command.

According to the IC, **E-35** from the nearby Sierra Fire District was the only mutual-aid company to respond to the fire. They were instructed to stage on Second Street just north of Lake Street and to stand by as a Rapid Intervention Team (RIT) for when search operations commenced.

During defensive operations, several units experienced water pressure problems as they operated master streams. **T-3** reported erratic water pressure, and a request was made through the Operations Officer to replace **E-10** that had been supplying their water. Command advised, however, that other units were experiencing the same problem, and that the water department had been asked to increase water pressure in the grid. Later-arriving units were instructed to use fire hydrants in adjacent water grids. These steps reportedly corrected the problem.

The IC then requested a special alarm for two additional engines, ladder truck, and heavy rescue squad.[13] These units were used to supplement ongoing defensive operations, establish alternative water sources in adjacent water grids, and assist with search and recovery operations. The units responding on the special alarm included **E-13, E-14, T-15, HR-11**, and a Safety Officer.[14]

T-15 was ordered to take a position at the northeast corner of the building and initiate aerial master-stream operations. **E-13** laid a 300-foot, 5-inch supply line from the hydrant at the corner of Evans in the rear of the building. The truck was instructed to prevent the fire from extending into the Annex and assist other aerial ladders in extinguishing the fire in the roof. **T-15** flowed at a rate of 800 gpm for an unknown period of time. The master stream was shut down temporarily during search and recovery operations. During this period a firefighter remained at the tip of the aerial watching for any flareups. **T-15** was repositioned at approximately 6 a.m. and replaced the 1-3/4-inch tip with a fog nozzle to wet down hot spots.

---

[13] The time at which the special alarm was requested and dispatched was not provided.

[14] The dispatch and response times for E-14 and HR-11 are believed to be inaccurate, and E-13 and T-15 were not listed on the incident report as responding to the fire.

due to the fluctuating water pressure, **E-14** reverse-laid a 500-foot, 5-inch large-diameter supply line from **E-5** to a hydrant at the corner of Plaza and Lake Streets. This hydrant was selected because it was located in a separate water grid and would not affect the water pressure of the other units involved with the ongoing defensive operations.

During the incident, three Safety Officers were used to monitor fireground activities. Two were deployed on Side A and one on Side C of the building. The Safety Officers had the authority to stop and report all unsafe fireground practices and activities to the IC. Despite the intense heat and smoke conditions inside the building and the partial collapse of the building, there were no firefighter injuries reported. Figure 8 shows the positioning of fire apparatus at the fire scene during the fire.

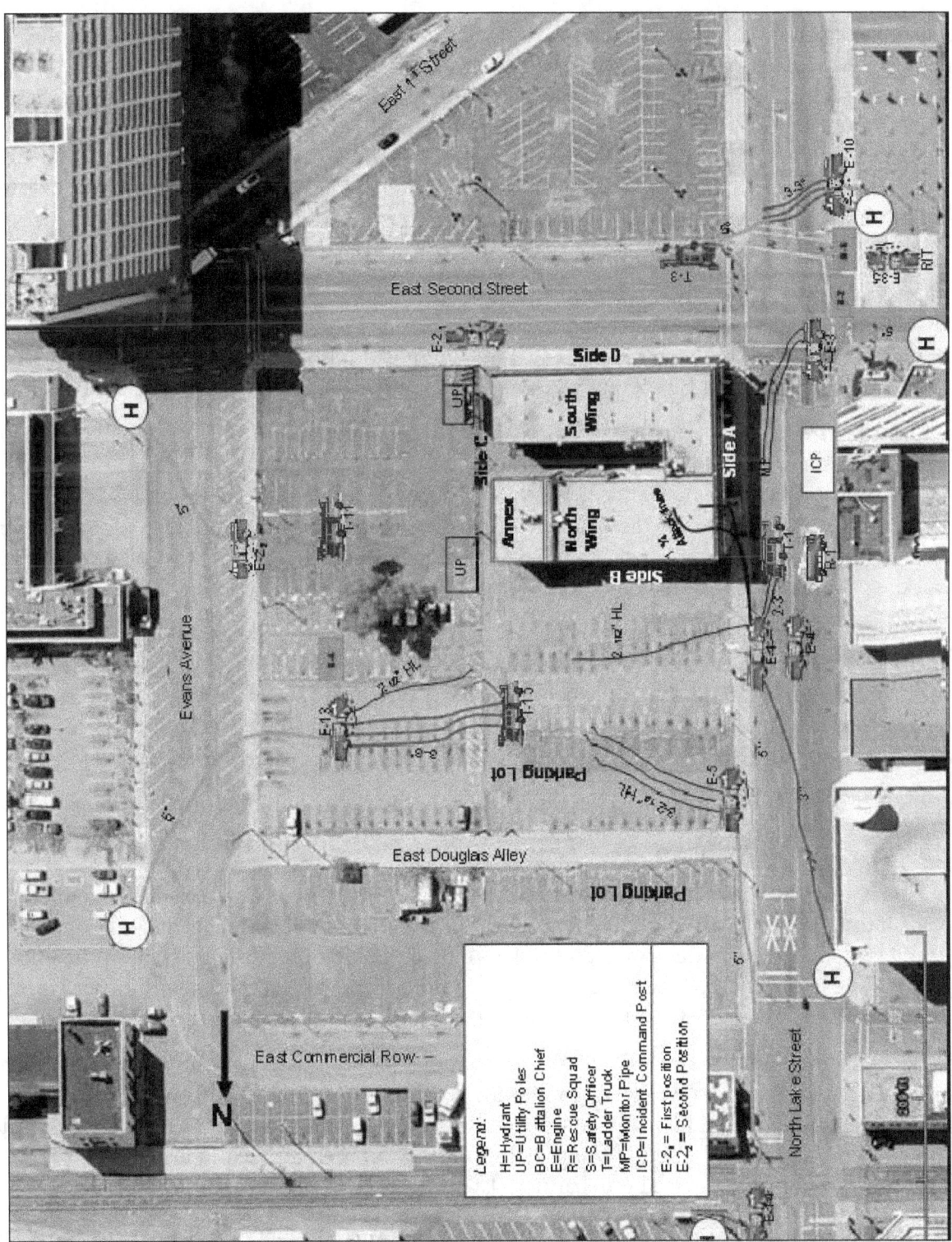

Photo, Courtesy of ATF

Figure 8. Positioning of Fire Apparatus.

SEARCH AND RECOVERY OPERATIONS

At the conclusion of defensive operations, search and rescue operations were initiated by fire and emergency service personnel. Generally, there are two components of the search and rescue process. The first is the primary search, which usually is conducted when fire suppression operations are initiated. Interior conditions usually are so threatening that the primary search involves a quick once-over of the entire accessible area, with an emphasis on areas where someone is most likely to be found. The secondary search is performed after the fire has been brought under control, and interior conditions have improved. The secondary search is a much slower and thorough process to locate all victims.

At the Mizpah fire, the rapidly spreading flames and questionable stability of the building forced the IC to withdraw personnel from the building before a primary search could be implemented fully. A thorough search of the building's interior was not attempted until the fire was brought under control. In this case, the secondary search was, in fact, the first search of the interior of the building.

Once master-stream operations had concluded, **E-10** and **T-3** were assigned to search the south wing. The south wing was searched first because the structural stability of the north wing was in question. Also, if there was a chance that someone survived the fire, the likelihood was greatest in the south wing. Before entering the structure, the Operations sector officer briefed the search teams. This was a good idea because each member performing search functions must have a clear idea about the parameters of the search.

The crews divided into two search teams and entered the south wing at the second floor level by way of **T-3's** aerial ladder. Conditions on the second floor were poor; visibility was limited due to thick smoke, and it was hot. The search teams conducted a systematic search of the second floor. **E-10** crew searched the interior rooms while **T-3's** crew searched the outside rooms that paralleled Second Street. They marked the doorframes with a large X showing the room had been searched.

During the search **E-10** discovered a deceased victim in the bathroom in Room 93. Command was notified of the find and the crew was instructed to mark the door, and not to move any victims. **T-3's** crew discovered an elderly man asleep in his room; apparently he had slept through the fire. He was immediately escorted down the aerial to safety. At this point, **E-10's** crew had to leave the building to replenish their breathing air. **T-3** continued to search of the second floor and found a second victim near the center staircase in the west hallway. They noted the location of the body and proceeded to the third floor and continued their search. The third floor was reported cooler, with better visibility. Eventually, the crew from **T-3** had to leave the building to replenish their air supply.

The crews from **E-4** and **E-14** were assigned to continue the search of the south wing. **T-3** briefed the replacement crews as to what they found, the conditions inside the south wing, and where to pick up the search. No additional fatalities were discovered by **E-4** and **E-14** search teams. After completing the search of the south wing, **E-14's** crew was ordered to extinguish any hot spots in the north wing. **T-3** was directed to rehab, and later released.

E-4 and **R-1** were instructed to search the Annex. **E-14** was assigned as the RIT. While the Annex was being searched, a firefighter atop the **T-15's** aerial ladder kept a watchful eye for any flareups of the fire. The search crews entered the Annex through a ground-floor door at the rear of the building. They commenced to search the first floor and reported conditions clear. As they advanced to the second floor the crew discovered a partial collapse of the third floor. The search was stopped immediately, and the crews were removed from the building. Once out of the building a PAR was ordered, and all personnel were present. Figure 9 shows the partial collapse of the north wing's third floor south hallway.

Figure 9. Overhead Photo of Collapsed Areas of Third Floor.

When **HR-11** arrived on the fire scene they were assigned to assess the structural stability of the Annex. The lower floors of the Annex were determined to be safe, and **HR-11** and **E-14** were assigned to complete the search of the Annex previously started by **E-4** and **R-1**.[15] During the search, two additional fatalities were found on the third floor; the rooms were marked and Command was advised. Search crews were unable to search the fourth floor because of unstable ventilation equipment on the roof. The search crews were relieved and sent to rehab to rest and recover. The following morning the RFD US&R Team reassessed the stability of the building and stabilized it with shoring.

---

[15] HR-11 position on the fireground was not provided by the fire department and therefore is not shown in Figure 8.

UNIFIED COMMAND

A primary objective of a Unified Command is to ensure mission integration and interoperability across functional and jurisdictional lines, as well as between public and private organizations. This was accomplished early in the incident. At the onset of the incident, the Battalion Chief quickly assumed Command and established an ICP. Incident Command transitioned from the traditional Incident Command System (ICS) to a multiagency Unified Command as more city departments and local and Federal agencies became involved. Unified Command enabled agencies with different legal, geographic, and functional responsibilities to coordinate, plan, and interact more effectively. During large-scale multiagency operations, a Unified Command is more efficient than agencies acting independently, and it helps eliminate duplication of effort. During the fire, officials from the different agencies worked together to formulate and execute the IAP. According to the IC, all decisions made at the Command level were based on a consensus opinion among the various agency representatives.

According to the IC, the Unified Command worked extremely well throughout all operations. The lead fire investigator with the ATF commented that this was the best managed and coordinated major fire incident that he had ever seen.

Agencies represented at the ICP during the incident included Reno fire and police officials, Reno's ATF Field Office, the ATF NRT and support group, REMSA officials, Public Information Officer (PIO), Washoe County District Attorney's Office, Washoe County Coroner's Office, Reno Building Department, Washoe County Sheriff's Office, American Red Cross, Nevada Occupational Safety and Health Administration (OSHA), and public utility officials.

The Command structure incorporated the major functions of Incident Command, Operations, Planning, Logistics, and Finance/Administration. Figure 10 shows the major operational components during the incident.

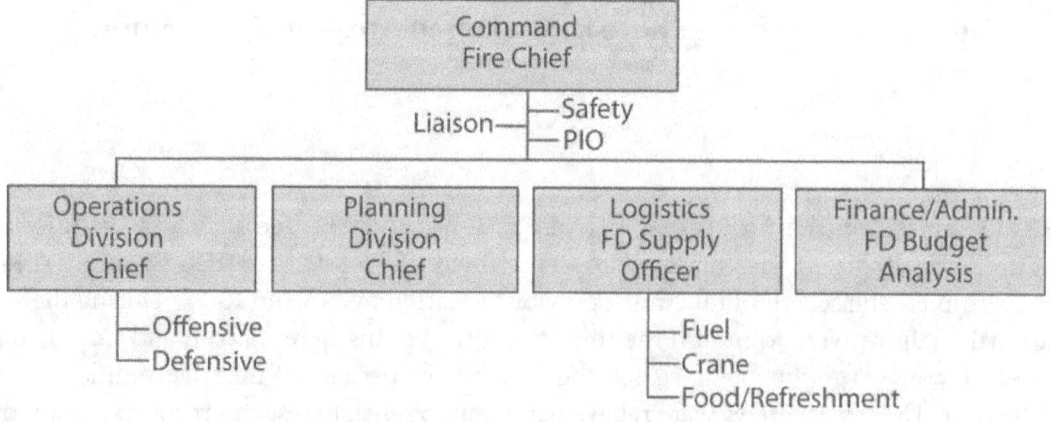

Figure 10. Incident Command Functions.

The incident was divided into offensive and defensive modes of operation. The primary goals of offensive operations were to locate, confine, and extinguish the fire within the building of origin, rescue all trapped building occupants, restore situational control, and restore normal activities.

The primary objectives of the defensive operations were to prevent the fire from extending into the south wing, account for all building occupants, maintain an adequate and reliable water supply, ensure fireground safety, and conduct a thorough postfire investigation.

According to the IC, offensive operations were divided into four primary branches: EMS, Law Enforcement, Safety, and Suppression. The Suppression Branch was further divided into a Rescue Group, Division 2 for second-floor operations, and Division 3 for third-floor operations. Figure 11 shows the different branches of operations and unit assignments during offensive operations. Figure 12 shows the defensive components during the incident.

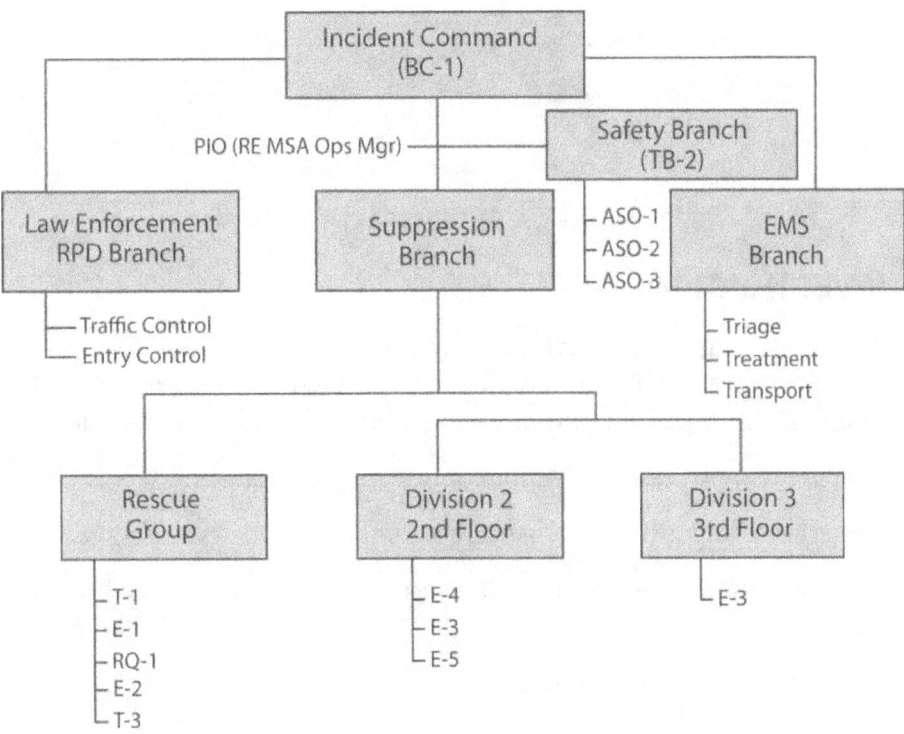

Figure 11. Offensive Operations.

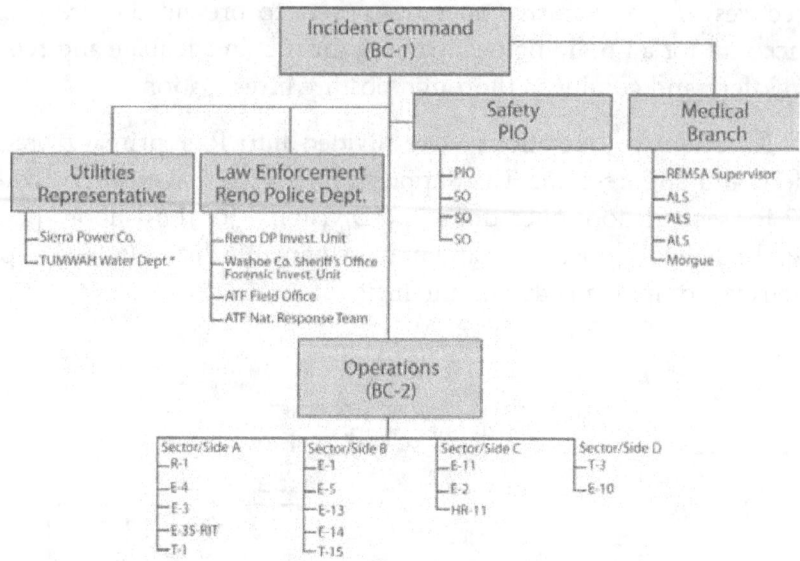

Figure 12. Defensive Operations.

POSTFIRE INVESTIGATION

Given the magnitude of the incident, local fire and police officials requested assistance from the ATF for determining the origin and cause of the fire. ATF activated the Western NRT. Approximately 18 special agents from the San Francisco field office joined the investigation. The NRT consisted of:

- a team leader, who was responsible for coordinating operations;

- a team supervisor responsible for managing the team and determining team assignments;

- at least one certified fire investigator (CFI) responsible for determining the origin and cause of the fire;

- an explosive enforcement officer (EEO) responsible for evaluating any explosive devices that may have been involved;

- a fire protection engineer (FPE) for evaluating the building design features, fire suppression systems, and analyzing the fire spread through computer modeling;

- a photographer for documenting the fire scene;

- a forensic chemist;

- an accelerant detection canine (ADC); and

- an evidence technician for collecting and documenting all evidence collected.

RFD investigators, assisted by ATF-NRT and the Reno Police Department and the Washoe County Sheriff's Office conducted an 8-day investigation into the origin and cause of this deadly fire. The Reno Police Department obtained a search warrant to permit fire investigators to enter the building and conduct an origin-and-cause investigation, and to maintain custody of the fire scene. Prior to initiating the postfire investigation, RFD and ATF fire investigators briefed the fire department IC on the postfire investigation's process. Figure 13 shows the organizational structure of the postfire investigation.

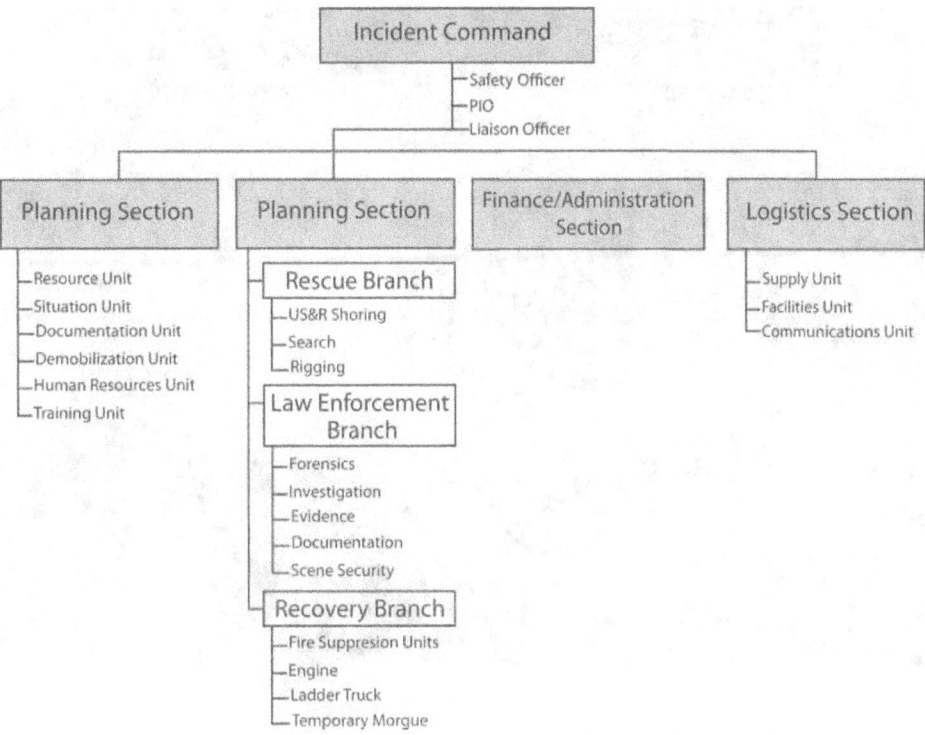

Figure 13. Components of Postfire Investigation.

ATF and police investigators interviewed several residents and employees of the hotel to learn what they knew about the fire and the conditions inside the building during the early stages of the fire. According to the hotel's chief maintenance man, the hotel manager had arranged to have a dumpster delivered on Thursday to pick up the old mattresses awaiting disposal, but the dumpster never arrived. Maintenance workers continued stacking the old mattresses along the walls of the hallways in the north wing. New mattresses were placed along the walls as well. Fire crews were interviewed to learn what they witnessed as they approached the scene and the conditions inside the building when they initially attacked the fire.

Fire Investigation Process–A systematic postfire investigation was conducted of the building, moving from areas of least fire damage to areas most heavily damaged. Scene processing began with an inspection of the exterior of the building. The entire exterior of the building was photographed from the ground, from overhead using aerial apparatus, and from the roof of Harrah's Hotel and Casino across the street from the Mizpah Hotel. It was determined that the fire had been contained to the north wing, and did not extend into the south wing, which was the goal of the

Photo, Courtesy of ATF

Figure 14. Extent of Fire Damage–North Wing.

Fire investigators then moved inside the building starting on the ground floor of the north wing and Annex. Since the north wing had sustained significant fire damage, members of the RFD US&R Team accompanied the investigators to ensure the building was structurally safe.

There was no significant smoke or fire damage to the first floor of the north wing or the adjoining Annex. These areas sustained mostly water and residual smoke damage. There was significant smoke and fire damage to the area at the top of the west staircase, indicating that smoke migrated from the second floor to the first. This was consistent with burn patterns observed by fire investigators that showed the fire migrated from the second floor by way of the western stairwell to the third floor.

The south wing sustained only minor damage, primarily from smoke. The heat and smoke patterns show that the heat and smoke spread to the south wing via a second-floor corridor connecting the two wings. The investigators were able to determine that the fire had not started in the south wing and ruled it out as the point of origin for the fire.

The focus of the fire investigation then shifted to the third floor of the Annex. Very little fire damage was found in this section of the building. However, while searching the third floor, the investigators discovered two fire victims: one in Room 58 and the second in Room 62.

It appeared that the smoke and heat may have migrated into this area of the hotel, as occupants were fleeing the building and leaving their room doors and windows open. The third-floor foyer of the Annex sustained substantial smoke and heat damage, but very little actual fire damage. In the rooms where the doors were found closed, there was little smoke or heat damage, which indicates that the migration of the smoke and heat may have been influenced by the open doors and windows. Photos taken by ATF during the fire investigation showed that the room doors were not equipped with self-closers, which would have closed the door automatically as the occupants left their room, slowing the spread of the heat and smoke. Today all hotel rooms are required to be equipped with self-closers.

Due to a partial collapse of the stairwell leading to the fourth floor of the Annex, investigators were forced to enter that floor via an aerial ladder through a window in Room 66. A fire victim was discovered in Room 69. Every room on the fourth floor showed some visible signs of smoke damage despite their doors being closed. Considerable heat and smoke migrated to the fourth floor Annex, but there was little to no actual fire damage. The failure of the skylight may have created a flue effect that caused the heat and smoke to move quickly into the fourth floor. No other victims were discovered during the search of the Annex.

Based on their observations of the Annex, fire investigators were able to rule out the Annex as the area of origin for this fire. All patterns throughout the Annex indicated that the fire migrated there from the north wing.

The following day, the investigation team began to examine the second floor of the north wing where three more fire victims were discovered. The first was found in the north hallway outside Room 7. After the body was removed, the investigation of the north hallway continued toward the front of the north wing.

A second investigation team began a simultaneous examination of the north wing south hallway. A second victim was found in the south hallway outside of Room 12. The victim's body was partially covered by remnants of two mattresses. A partial collapse of the third-floor was also discovered above the south hallway approximately 20 feet west of the east stairwell. Following the discovery of the third-floor partial collapse, the investigation of the north wing was stopped temporarily until this area was secured by the department's US&R team.

The investigation team working the north hallway then moved to the third-floor via the eastern stairwell stair. The fire appeared to have spread up the staircase to the third-floor. After clearing the east staircase, investigators returned to the west end of the south hallway where they found a third victim outside the entrance to Room 16. The east end of the south hallway had sustained heavy fire damage, which was consistent with the location of numerous mattresses. The heavy fire damage in this area was attributed to the fuel load from the mattresses.

The investigation team then moved to the west staircase. The investigative team entered the staircase at the first-floor level. The stairwell sustained heavy smoke and fire damage at the second-floor landing. The fire damage was more pronounced to the left (north side), and less to the right (south side). The fire damage described here was consistent with the intensity of the fire described by the fire crews from **E-4** and **E-3** who were assigned to attack the fire during offensive fire operations. Burn patterns showed that the fire spread from the north to south, and west to east. The central staircase door in the west hallway showed signs of heavy fire impingement. Fortunately, the stairwell door held.

At this point, the Reno Police received word that the person who lived in Room 21 may have been connected to the fire. The RPD secured a search warrant for the resident's room. An accelerant detection canine also was brought in to check for the presence of accelerant/flammable liquids, but none was found. Rooms 21 and 22 were examined and were found to have sustained light fire damage. Another fire victim was located near a small alcove across from Room 79.

While waiting for the third floor of the north wing to be stabilized, two more fire victims were observed from an overhead basket suspended from a crane. The first victim was found at the eastern end of the north hallway, near Room 31. The second victim was located at the east end of the south hallway, near Room 34. The twelfth and final victim was found outside Room 35 in the south hallway on the third floor. Two cadaver dogs were summoned to search the second and third floors at both ends of the collapse zone. The dogs alerted at both ends of the collapse zones, but no additional bodies were found.

According to the IC, many of the building occupants who attempted to exit the building through the interior corridor perished in the fire. Seven of the 12 fire fatalities were found in the hallways during the postfire investigation. Figure 15 and Figure 16 show the locations where the fire victims were found.

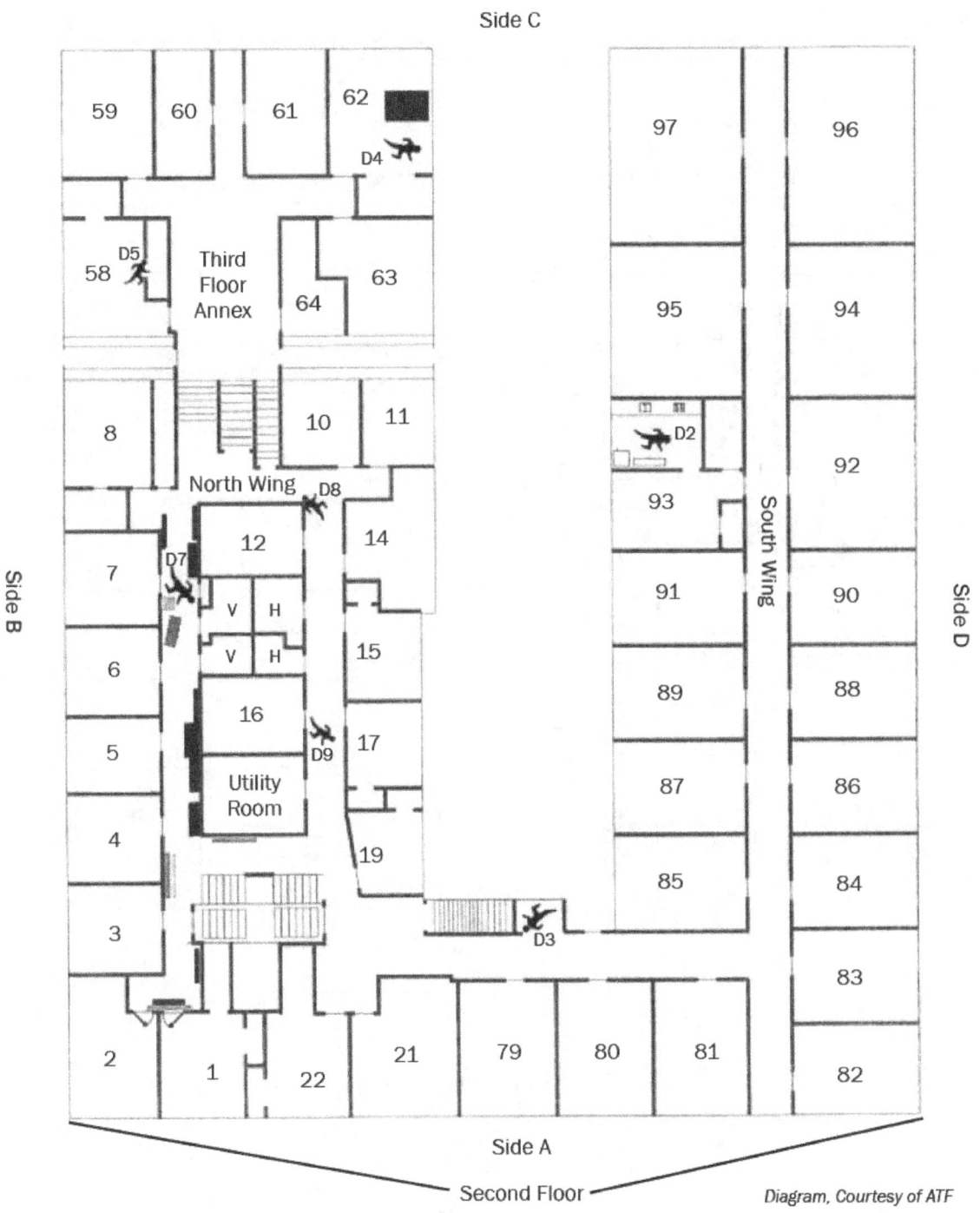

Figure 15. Fire Victim Locations: Second Floor and Third Floor Annex.

Side C

Side B

Side D

Side A

Third Floor

Diagram, Courtesy of ATF

Figure 16. Fire Victim Locations: Third Floor and Fourth Floor Annex.

Evidence of a "V" pattern was found above remnants of mattresses and box springs along the wall near the western stairwell. The "V" pattern suggested a localized plume. The pattern was caused by the increased fuel load from the mattresses. Similar patterns were found along the entire north hallway, each corresponding to the mattresses lying along the floor.

The door to the western stairwell was examined and found to be detached from the upper hinges, causing the door to tilt eastward. The fire had consumed the upper third of the door closest to Room 1. This suggested that the fire came from the west, which was the foyer area in front of Rooms 1 and 2.

Two mattresses were found in the foyer. Based on their footprint, the investigators were able to determine that the mattresses were in front of the door to Room 1. The fire pattern indicates that the fire started in the foyer in front of Rooms 1 and 2. The patterns also indicate the fire was moving away from the foyer throughout the rest of the building (west to east), which is consistent with observations made by the fire crews attacking the fire on the second floor.

The burn patterns along the north and south hallways of the north wing were representative of the 16 mattresses and box springs found along the second-floor hallways. Figure 17 shows the remnants of mattresses along the second-floor hallway. Some mattresses had been stacked together three deep. The mattresses supplied the fuel for the rapidly spreading fire through the hallways.

Photo, Courtesy of ATF

Figure 17. Remnants of Mattresses in the Hallway of the Second Floor.

The fire spread was verified by faults in the electrical distribution system. By following the electrical distribution lines down the two hallways on the second floor, the investigators were able to find two separate arcs. The first was located in the north hallway near Room 5, and the second was in the southern hallway near Room 19. Since the main distribution panel and power was coming from the east, these arcs help indicate the progression of the fire. The arc in front of Room 5 indicated the fire came from the west, and the arc near Room 19 indicated the fire came from the north. Through arc fault mapping, the electrical engineer was able to verify that the fire originated in the northwest corner of the second floor.

Eyewitness Accounts–There were two eyewitnesses to the incipient stage of the fire. With the physical investigation completed, the eyewitnesses could help to verify the point of origin.

- One resident of the hotel arrived at the hotel and entered the building through the front door. He walked up the west staircase to the second floor, and proceeded to the north hallway. At the top of the steps he had to avoid running into the woman who was alleged to have started the fire. He returned to his room and turned on the TV. Almost immediately, the fire alarm activated. He contacted the front desk and they were aware of the alarm activation. He then exited his room and saw light smoke at the ceiling level. As he walked west toward the second floor of the north wing the smoke became very heavy and dark. The smoke became so thick he could not see and he fell down the steps. As he attempted to stand up he said the smoke was incredibly hot above his head. He crawled up the steps and exited by way of the fire escape.

- At approximately 10 p.m., a hotel resident was returning from the bathroom on the second floor when he noticed the alleged firesetter standing by several mattresses in the hallway near Room 1. Shortly after entering his room he heard the fire alarm activate. He remained in his room until he saw smoke coming under his door. He then opened his door and looked down the hallway and saw what he described as two or three mattresses on fire near Room 1. The suspect was gone. The resident attempted to exit the building using the west staircase nearest the fire, but was driven back by the intense heat and smoke. He was forced to exit through the rear of the building. After escaping he walked around to the front of the building and saw a woman hanging out of a window on the second floor yelling for help.

Based on the fire patterns found throughout the building and statements from the fire crews and witnesses, and elimination of other sources of ignition, it was the opinion of the fire investigators that the fire began in the alcove on the second floor of the north wing in the mattresses that were placed in front of the doors of Rooms 1 and 2. The fire was intentionally started by an open flame applied to the mattresses.

FIRE LOAD AND SPREAD

Fuel Load–Several factors influenced the rapid spread of the fire. The first was fuel load. There were mattresses in the north hallway the day of the fire; the bulk of those old mattresses were ready for removal from the hotel. These were all twin institutional-grade mattresses and box springs, which measured 52 inches wide and 74 inches tall. The mattress tags read that the mattresses were 100-percent polyurethane foam, having a net weight of seven pounds of filling. Figure 18 is a photo of a mattress tag taken by ATF agents; it lists the type and quantity of padding contained in the mattresses that were stacked along the hallway walls.

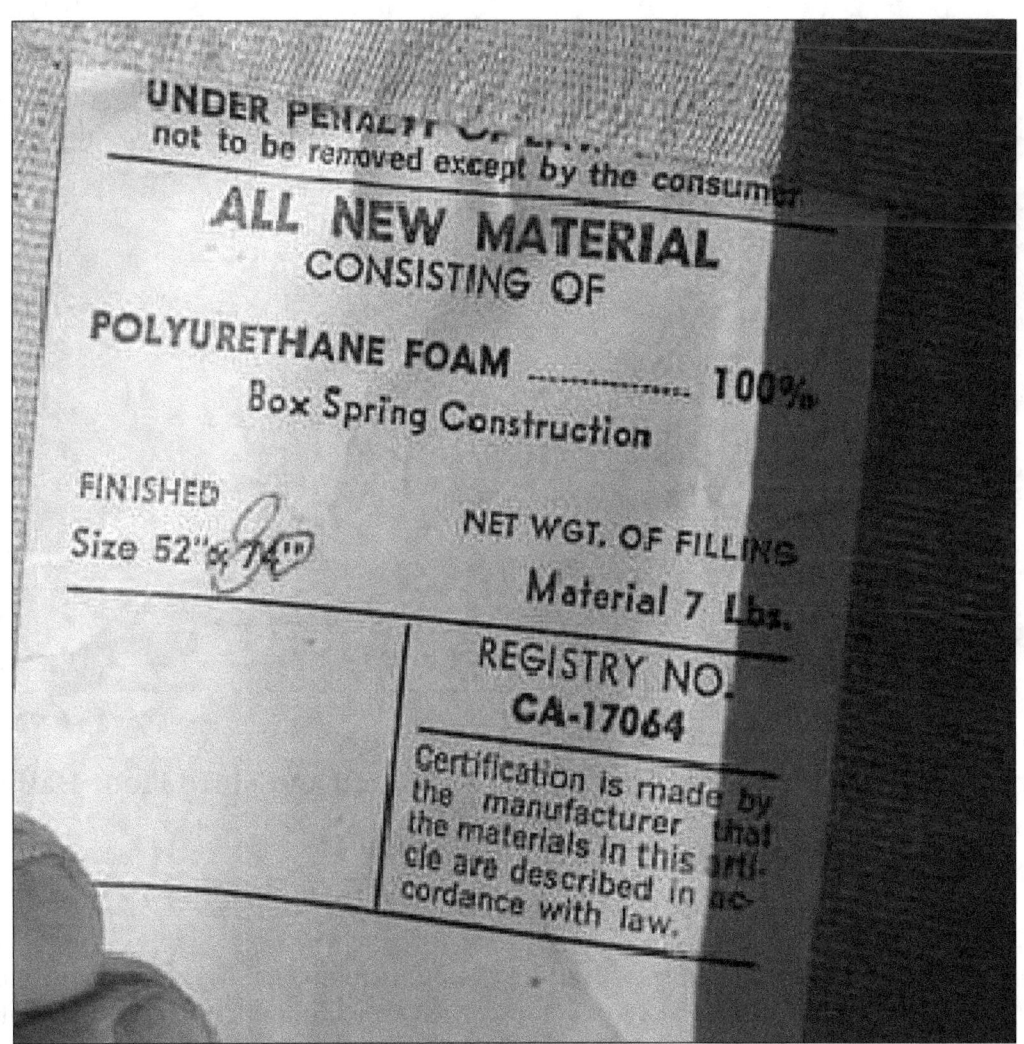

Photo, Courtesy of ATF

Figure 17. Remnants of Mattresses in the Hallway of the Second Floor.

Polyurethane foam burns if exposed to a sufficient heat source. The melting point of polyurethane foam is 350 to 375 °F (177 to 191 °C), and it ignites at temperatures above 500 °F (260 °C). Once ignited, the polyurethane foam can cause rapid fire spread, intense heat, and dense black smoke.

Polyurethane foam has toxic gases. Flexible polyurethane foam such as that found in mattresses, once ignited, may degrade and melt into a combustible liquid, which adds to the fire involvement.[16]

The manner in which the mattresses were stacked against the wall aided in the spread of the fire. The mattresses were stacked vertically, and in some cases horizontally against the hallway wall. This placement allowed air to circulate around the mattresses, rapidly spreading the fire. Figure 19 shows the typical mattresses and how they were stacked along the walls of the hallways.

Photo, Courtesy of ATF

Figure 19. Typical Mattresses That Lined the Second and Third Floor Hallways.

According to the IC, the accelerated spread of the fire on the second floor of the north wing was believed to have been aided by a strong cross-draft created by windows, doors, and a skylight open at the rear of the building on the second floor, and the open staircase at the front of the building.

Based on their findings, fire investigators formulated a hypothesis as to how the fire spread. A fire protection engineer with ATF developed a field computer fire model of the fire. The fire model placed the initial fire in the second-floor foyer in front of Rooms 1 and 2. When the model ran it replicated the fire just as the investigators had envisioned it. Although not all of the variables were placed in the model, the model showed a fast-moving fire with the fire spreading throughout the

---

[16] National Fire Protection Association. *Fire Protection Handbook*. 19th Edition, Vol. II, pp. 8-165, 8-170.

second-floor north wing of the building. The fire then simultaneously advanced up the eastern and western stairwells and extended into the third-floor area. Smoke and heat began advancing up the stairs to the third floor within seconds. The actual fire did not extend to the third floor for about 2 minutes. The rapidly spreading smoke and heat up the eastern and western stairwells would explain the cluster of fire victims found in the hallways on the third floor. According to the fire protection engineer, the enormous fuel load, extreme heat, and rapid extension of the fire precluded a successful offensive fire operation from the outset.

The costs associated with retrofitting the hotel with sprinklers in all probability would have been much less than the cost associated with rebuilding the north wing and repairing the smoke and heat damage to the south wing and Annex.

EMERGENCY MEDICAL SERVICES

EMS in Reno are provided using a partnership, with RFD providing first response and REMSA providing EMS transportation. Governmental oversight of EMS is provided by the Washoe County District Board of Health which promulgates regulations, oversees EMS provision, and coordinate mass casualty incident (MCI) response for Reno, Sparks, and other areas of Washoe County.[17]

RENO FIRE DEPARTMENT

The department provides EMS as a nontransporting first responder for the city. Reno firefighters are all EMT-Basic (EMT-B) certified with 120 certified as EMT-Intermediates (EMT-I). Twelve of Reno's 18 engine companies provide EMT-I-level care. Due to the intense demand for rescue and fire suppression activities at this incident, firefighters provided little emergency care beyond that provided during the initial rescue.

Officials from both RFD and the REMSA believe that interagency relations are very good. The roles of each agency are well-defined and RFD believes that their current EMS role is appropriate.

REGIONAL EMERGENCY MEDICAL SERVICES AUTHORITY

REMSA is a public utility (501c3) model that provides emergency and nonemergency medical transportation for Washoe County. Their territory includes the cities of Reno, Truckee, Gardnerville, and several unincorporated areas. REMSA has 35 ambulances, all ALS-capable. Each unit has a crew of either one EMT-Paramedic (EMT-P)/one EMT-I, two EMT-Ps or two EMS Registered Nurses (EMSRNs) who also can fill paramedic slots. REMSA responds to over 50,000 calls annually.

REMSA also fields three EMS aeromedical helicopters, one each in Reno, Truckee, and Gardnerville. These units serve mostly in rural areas and are staffed by one EMSRN and one critical-care EMT-P (CCEMT-P). The pilots are contracted from an outside agency. Critical care ground transportation is provided by a critical care ALS unit that is staffed by one EMSRN and one CCEMT-P.

Support units include nine "MedExpress" vans, each staffed by one EMT-B. These units provide medical transportation for those not needing traditional ambulance service (wheelchair van). One or two EMT-Bs work daily as Vehicle Support Technicians (VSTs). These personnel assist field units with restocking, light maintenance, and other support functions. Another support area includes special events (high-risk events, potential MCIs and other contracted events). These events are staffed by all levels of REMSA providers including physicians. EMT-Bs only provide patient care during special events or while working on MedExpress units. EMT-I is the minimal certification level needed to work an EMS unit.

Two EMS provider levels are uncommon for many EMS systems. EMSRNs are registered nurses, licensed by the State. They usually work on EMS units that handle interfacility critical-care transports. To work on the helicopter, REMSA requires 2 years of in-hospital ICU experience. CCEMT-Ps are critical-care paramedics who also are required to work with EMSRNs on helicopters and interfacility critical-care units.

---

[17] DHD. (December, 2005). *Washoe County District Board of Health Multi-Casualty Incident Plan.* (Revised, December 1, 2005), Washoe County, NV.

REMSA has a three-level management team including directors, managers, and supervisors. Under the vice-president of operations, directors oversee major branches of the organization including Deployment, Health and Safety, Human Resources, Finances, Aeromedical, Education and CQI, and Ground Operations. Managers focus on specific areas of major branches, while supervisors oversee smaller, specific areas, mostly in operations or training (Figure 20).

Medical direction is provided by a medical doctor who is a REMSA employee and works 4 days a week at REMSA. The doctor also practices emergency medicine at community hospitals. He is a nationally recognized EMS physician who has extensive involvement in EMS education. The emergency medicine community is represented by a group of local physicians: the Physician Medical Advisory Committee (PMAC). Included in PMAC are representatives from each community's emergency department who receive input from the medical community and review medical protocols.

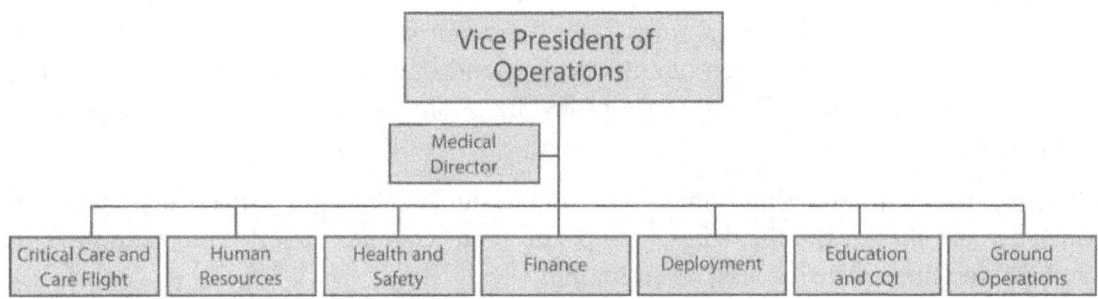

Figure 20. REMSA Officials.

Operationally, REMSA has a staff of 150 field personnel, mostly certified at the EMT-P or EMT-I level. All personnel have current State certifications and have been trained in the National Incident Management System (NIMS) to the IS-700, ICS-100, and ICS-200 levels.[18] REMSA's dispatch model uses system status management to determine how many and where units are placed. They usually are assigned to an area instead of a specific dispatch point (station). During evening shifts, 11 medic units usually are available for service. Personnel work 12-, 16-, or 24-hour shifts, with the additional ability to staff peak-load units.

They also provide a single dispatch system including a Medical Priority Dispatch (MPD) program. In 2004, REMSA's MPD program received a 3-year re-accreditation from the National Academy of Emergency Medical Dispatch.[19]

The night of the fire, REMSA was staffed at its normal strength of eight medic units plus two additional units placed in service due to Halloween activities. There were one operations supervisor, four dispatchers, one health-hotline operator, and two vehicle service technicians on duty. There were three fully-staffed EMS helicopters. Primary oncall managers were the EMS Special Operations

---

[18] REMSA. (2006). *Employees' Certification/License Status Report*, (11/29/2006).

[19] REMSA. (2006). *Employees' Certification/License Status Report, EMD*, (11/29/2006).

Manager, the EMS Operations Manager, and the Director of Ground Operations. Other staff personnel were available on backup call.

Washoe County has several personnel on call, including the Washoe County EMS Coordinator, Emergency Manager, and the Washoe County Health Officer. Washoe County also has an oncall search-and-rescue (HASTY) team available for response. The Washoe County Board of Health Multi-Casualty Incident Plan provides for mutual-aid response at the request of the IC.[20]

Emergency Medical Services Response to the Mizpah Incident

The following information outlines the response and actions of REMSA personnel. The EMS portion of the incident is described in three phases: notification and response, medical branch, and postincident evaluation.

Notification–REMSA became aware of the incident as fire units were responding. At 10:05 p.m. REMSA dispatch contacted RFD dispatch to determine if REMSA was needed. RFD advised that there were several injuries and requested "several units." Fire dispatch was not specific about the number of patients. REMSA decided to dispatch one medic unit and one operations supervisor (Medic 421, 175 [supervisor]). Supervisor 175 also was notified by phone. REMSA dispatch began planning for an MCI by transferring rural units into the city.

At 10:07 p.m., RFD requested three ambulances and designated a Staging Area. The EMS supervisor was contacted and moved the incident to a separate radio channel. Shortly thereafter, RFD called again and advised there were "at least seven injured."

Review of the REMSA tapes indicate that there are two main issues to address. First, the initial notification was actually from REMSA to RFD asking about the situation. Fire dispatch advised that EMS response was needed, but that advice lacked a sense of urgency or specific response directions. REMSA's decision to dispatch a medic unit and an EMS supervisor was appropriate, considering the information available. The second issue involved RFD's specific request for three medic units, based on information about "at least seven injuries." REMSA should have dispatched three EMS units instead of requesting direction from an EMS supervisor.

Medical Branch–The EMS supervisor was notified of the need for additional units just as Medic 421 was arriving on scene. A paramedic assumed command of EMS, and established an initial triage and treatment unit. Supervisor 175 requested a sizeup before requesting additional units. At 10:08 p.m., RFD advised that people were "jumping out of the building." Supervisor 175 requested one additional medic unit.

At 10:11 p.m., the Treatment Unit reported six minor injuries and the EMS supervisor advised REMSA not to dispatch any further units. Supervisor 175 arrived at 10:15 p.m. The paramedic from Medic 421 already had reported to the RFD IC. After conferring with the IC, Supervisor 175 became the Medical Branch Director.

Medic 421 advised there was one patient in cardiac arrest, possibly dead on arrival and five minor injuries. IC advised there were many trapped. Supervisor 175 took the following actions: Triage and Treatment Units were established formally on North Lake Street in front of the building, REMSA

---

[20] DHD. (December, 2005). *Washoe County District Board of Health Multi-Casualty Incident Plan.* (Revised, December 1, 2005), p. 7, Washoe County, NV.

dispatch was requested to initiate the MCI plan, a third medic unit was requested and a personnel all-call was initiated. At 10:22 p.m., Medic 424 was dispatched as "third-due unit." RFD advised REMSA of additional patients being moved. At 10:23 p.m., REMSA notified area hospitals that the MCI plan was in effect. One paramedic assumed the Triage Unit and another paramedic assumed the Treatment and Transportation Unit (Figure 21).

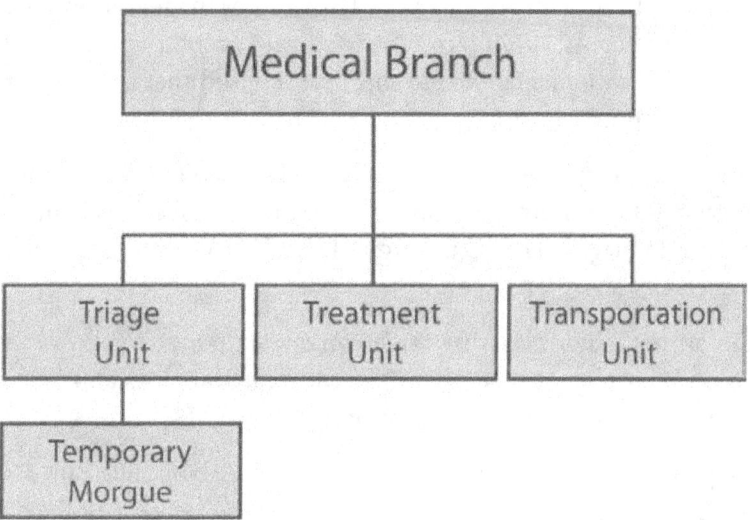

Figure 21. Incident EMS Branch.

At 10:25 p.m., dispatch started to notify management personnel of the incident. The Dispatch Manager was asked to respond to dispatch. One extra crew was recalled. The Operations Manager was notified but not asked to respond. The Special Operations Manager was notified and asked to take over REMSA staff notifications.

At 10:26 p.m., 175 provided an update of seven "green," and one "black."[21] The Medical Branch Director advised that no further units were needed and that one unstable patient was being brought down the ladder. At 10:29 p.m., REMSA updated area hospitals on the incident. RFD advised REMSA of additional patients at Second and Lake (unknown number).

At 10:30 p.m., REMSA dispatch recalled the REMSA Dispatch Manager and advised that the situation was "stable." Fire also advised that at Second and Lake (south side) there were additional smoke inhalation patients. RFD also advised that it had suspended providing medical first response to the city until the incident was stabilized. RFD would respond to specific EMS incidents as requested.

At 10:34 p.m., a patient update was provided: 18 "greens" and 1 "black" were on the north side and 1 red was being brought down the ladder. Treatment requested that the "red" patient be transported immediately. EMS units on scene included 175, 419, 421, and 424. The Washoe County Emergency Manager and the County EMS Coordinator were notified of the MCI.

At 10:36 p.m., the Medical Branch requested a fourth medic unit, and Medic 420 was dispatched.

---

[21] Standard EMS patient triage designations include Red = Critical, Yellow = Moderate, Green = Minor, and Black = Deceased.

REMSA also provided area hospitals with an update. At 10:37, the first "red" patient was transported to RENOWN (Reno Trauma) by Medic 424. One "yellow" patient was readied for transport. The patient status update was 3 "yellows" and 27 "greens" in the treatment area.

At 10:44 p.m., Operations Manager responded to the incident. Hospitals were updated on patient counts. Dispatch questioned the Medical Branch as to whether a large vehicle was needed to transport "green" patients. At 10:46 p.m., the REMSA Dispatch Manager assumed command of REMSA dispatch and was on the radio. At 10:50 p.m., the Medical Branch requested buses for shelter and to treat and transport "green" patients. The treatment unit advised that two city buses were already on scene. Patients were loaded onto buses.

At 10:53 p.m., Medic 420 transported five "yellows" to RENOWN (two medical and three trauma). The Operations Manager (173) arrived on scene and was assigned originally as the Transportation Unit Leader and Staging Manager. The IC reassigned the REMSA Operations Chief as the PIO. Extra blankets were requested.

At 11:03 p.m., the Treatment Unit reclassified one "green" patient to "yellow." At 11:06, the Medical Branch updated REMSA Dispatch. At 11:07 p.m., the Medical Branch established a temporary morgue. Nine "green" patients still needed hospital transport.

At 11:16 p.m., three patients were transported by 419 and one by police car to Saint Mary's Hospital. At 11:21 p.m., the Operations Manager (125) arrived on scene. Another "green" patient was upgraded to "yellow."

At 11:21 p.m., a fifth ambulance (425) arrived at the scene with supplies. At 11:24 p.m., Saint Mary's Hospital contacted REMSA dispatch with a rumor that all patients had been transported. REMSA clarified that patients were still being treated. At 11:27, Medical Branch requested the sixth ambulance to assist RFD with rehab; 419 advised that they were blocked by large-diameter firehose. At 11:33 p.m., 427 was dispatched to Second and Lake for rehab.

At 11:35, the Treatment Unit was advised to transport the next four "green" patients to Saint Mary's. At 11:37 p.m., 419 was able to transport these patients.

At 11:52, REMSA notified the hospital/s of patient updates. At about 1 a.m., the MCI was terminated and REMSA notified area hospitals.

In total, 17 REMSA personnel responded to the incident excluding those reporting to dispatch or for standby.

FINDINGS AND RECOMMENDATIONS

FIRE DEPARTMENT

Findings

The fire department's response and performance during the fire resulted in several positive decisions and outcomes, and areas needing improvement, which are listed here.

1. Rescue is the highest priority during a fire. Confronted with multiple rescues, the first-arriving units lacked a sufficient number of emergency personnel to conduct simultaneous attack on the fire and rescue operations. The IC made the correct decision to delay offensive fire operations and commit all available resources to effect the rescue of the trapped building occupants seeking rescue from their windows. Ground and aerial ladders, and utility trucks' "cherry pickers" summoned to the fire scene provided emergency service personnel the safest and most direct access to the trapped building occupants. This was the only way to reach the trapped occupants without hoseline support. To expedite the removal of the trapped occupants, the rescues were prioritized: those in the greatest danger were rescued first, and so on. Over 70 people were removed from the building by emergency service and civilian personnel. Over 30 rescues were made over ground and aerial ladders.

2. The IC and first-arriving unit officers used textbook positioning of aerial ladders that maximized the coverage along the front and sides of the building. Overhead powerlines at the rear of the hotel compromised the effectiveness of the aerial ladders there. Firefighters quickly requested bucket trucks, "cherry pickers" that could maneuver easily around the overhead powerlines and effect the rescue of trapped occupants in the rear of the building.

3. An ICS was established quickly and later transitioned smoothly to a Unified Command as other local, State, and Federal agencies became involved in the incident. The Unified Command ensured mission integration and interoperability across functional and jurisdictional lines as well as between public and private organizations. Operational efficiency also was improved by eliminating duplication of effort among the many public and private agencies involved in the incident. Agency officials worked together through the Unified Command to formulate and execute the IAP, and were able to convey and coordinate the strategic and tactical plan effectively with all forces involved in the incident.

4. Care was taken to ensure firefighter safety in a very hazardous situation. No injuries were reported to emergency service personnel during the fire. The fire department quickly initiated its personnel accountability system. Several PARs were ordered by the IC during the active and postfire operations to account for personnel. RITs were deployed and stood ready to intervene if necessary. Three Safety Officers were deployed on the fireground to monitor tactical operations, and had the authority to stop any unsafe operations immediately. A rehab area was established where emergency crews could rest and recover.

5. **Lack of sprinklers.** A major factor that contributed to the rapid fire spread was that the hotel was not equipped with an automatic sprinkler system. Had the hotel been equipped with automatic sprinklers, there is little doubt the fire could have been contained or even extinguished prior to the fire department's arrival. Most, if not all, of the fire fatalities might have been avoided if the building had been sprinkled.

Recommendation: Local governments should review their local building code and ordinances and eliminate or reduce costly requirements, such as connection fees and high water rates, that often serve as disincentives to property owners in retrofitting existing high-hazard occupancies with sprinkler systems. Property owners should be encouraged to install sprinklers rather than discouraged by such fees or restrictions.

6. **Foam mattresses.** Another factor that contributed to the tragic fire was the numerous polyure-thane foam mattresses that were stacked along the second- and third-floor hallways for several days prior to the fire.

    Recommendation: As hotel personnel replace old mattresses, the old mattresses should be removed from the building immediately and disposed of properly. Local governments should consider adopting local ordinances that would mandate the local fire department to be notified of such occurrences, and to ensure that all bedding has been removed from the building and properly disposed of.

7. **Rapid water supply.** Despite evidence of heavy smoke conditions prior to leaving the fire station, the first-arriving engine failed to establish an adequate and reliable water supply, and had to rely on tank water until a later-arriving engine could lay a supply line from a nearby hydrant to the first-arriving engine. Although it may have taken a minute or so for the first-arriving engine to connect to the fire hydrant, there would have been little impact on rescue operations, and the engine company would have established a reliable water source to initiate an attack on the fire.

    Recommendation: Fire departments should adopt Standard Operating Procedures (SOPs) that designate apparatus positioning and tactical responsibilities for residential and commercial building fires. Such procedures provide a structured yet flexible framework for standard tactical operations and deployment of personnel and fire apparatus. Tactical responsibilities such as water supply and interior assignments should be incorporated into the building's preplan, and disseminated to all fire companies that normally would respond to a fire in the subject building.

8. **Water pressure.** As more master streams were deployed to extinguish the fire, several engine companies began to experience significant fluctuations in water pressure. The fluctuating water pressure was the result of large water demand created by the numerous master streams, and the inability of the domestic water system to keep pace with the demand. The situation was corrected by requesting the city water department to increase water pressure on the affected water grid, and by directing later-arriving fire units to fire hydrants in a different water grid.

    Recommendation: Good preplanning can help reduce the chance of this happening during a major fire. Once the amount of water necessary to extinguish a fire in a structure has been determined, the fire department should conduct a flow test of the domestic water system to determine if it is capable of meeting the fire flow demand. If the system is incapable of meeting the flow demand, the fire department should contact the local water department for possible solutions to the problem. It may not be possible to increase water pressure on a particular water grid because the system may not be capable of handling the added pressure. Large cities that have aging infrastructure may be incapable of meeting the fire flow demand. It is best to determine before the fire if the water system can meet the challenge during a major fire. If not, an alternative water supply to meet the flow demand must be identified. In such cases, the preplan should identify alternative water supplies, i.e., an alternate water grid, static water supplies such as lakes, ponds, or rivers, and water tankers.

Fire departments should establish a Water Supply Officer position. This individual would respond on all multiple-alarm fires or when requested by an IC. This individual should have a thorough knowledge of the city domestic water system and maps of the different water grids and control valves. In areas without a water distribution system, the Water Supply Officer should develop fire flow calculations during preplanning for all major and high-risk structures and identify water sources that could be used to support tanker operations.

9. **Interior attack lines.** Firefighters initially attacked the fire using a highrise pack consisting of a 3-inch feeder line that supplied two 100-foot 1-3/4-inch handlines. Based on the interview testimony from the officers assigned to attack the fire it is doubtful that two 1-3/4-inch attack lines would have delivered a sufficient water flow to extinguish the fire. In addition, the 100-foot attack lines were not long enough to cover the entire second floor of the north wing.

Recommendation: This situation normally is addressed in the building preplan. Preplans should consider the worse-case fire scenario, and take into consideration the type, size, and construction type of the occupancy under consideration. The buildings height, length of staircases, and hallways should be key considerations when determining the length and diameter of hoselines to deploy during a fire.

EMERGENCY MEDICAL SERVICES

Findings

REMSA's response and performance resulted in several positive outcomes. All patients who were injured but still alive upon EMS arrival survived the incident. There were no significant injuries suffered by EMS personnel.

Several decisions and actions taken by REMSA likely contributed to the success of the operation:

1. **IC.** REMSA personnel functioned well under the ICS. Most importantly, the first EMS official immediately reported to the IC and began to set up a Medical Branch and began patient triage. Transitions of Command were conducted properly and included a face-to-face meeting between the IC and REMSA supervisors.

2. **MCI.** Early in the incident, the EMS Supervisor instructed REMSA dispatch to initiate the MCI plan. This allowed REMSA to augment EMS forces quickly and ensure that the remainder of their response area was covered adequately. All patients were triaged and provided with triage tags.

3. **Management official.** A limited number of high-ranking management personnel responded directly to the incident. This prevented the EMS Branch from becoming "top heavy" and allowed REMSA senior management to concentrate on long-term matters. It also helped to ensure that senior managers were rested and ready for service in case of a long-term incident. Those senior managers who arrived on scene, allowed the REMSA supervisor to continue in command of the Medical Branch. REMSA senior managers were able to fill other ICS positions and act as advisors to supervisors.

4. Early activation of the trauma communications system allowed local hospitals and specialty centers to be ready for surge capacity. City services quickly provided large buses for delayed treatment and shelter. Since an accurate patient count was not available, early notification was critical for specialty centers (burn, hyperbaric) because there was a possibility that they would receive many critical patients.

5. **Shelters.** The local American Red Cross chapter responded quickly and opened a high school to provide a shelter for residents displaced from the Mizpah Hotel. Along with management personnel, the team included a family practice physician who provided assessment and care for several Mizpah Hotel residents suffering from chronic medical conditions.

6. Patient care was appropriate, and those who were injured were transported to an appropriate hospital.

7. A review of records indicated that all REMSA personnel were certified and licensed at the appropriate level. All personnel had completed NIMS training including IS-700 and ICS-100. Supervisory personnel were trained in ICS-200.

8. Standard patient tracking forms and patient care reports were completed properly. Documentation was completed as per the Washoe County MCI Plan.[22]

9. An internal critique was conducted 17 days postincident. REMSA strengths and weaknesses were identified, some of them mentioned above.

10. Overall, EMS was provided effectively during the Mizpah incident. EMS personnel were challenged with a significant incident where it was impossible to determine the extent of the devastation quickly. Further, firefighter first responders were unavailable to provide first responder assistance, as they were committed to fire suppression operations.

The following recommendations are provided for training and evaluation purposes and were discussed with REMSA prior to the release of this report. Overall, REMSA personnel performed admirably during the entire incident. The recommendations that follow would not have changed patient outcomes.

1. **Recommendation:** Upon determining the possibility of a major MCI, a predetermined response should have been dispatched. In this case, REMSA should have dispatched an EMS Strike Team to the scene.[23] Unnecessary time was spent determining what units to send. Dispatch should not have to contact a field supervisor to initiate an MCI response.

2. **Recommendation:** EMS supervisors should avoid giving radio instructions before arriving at the scene. Doing so violates the integrity of the ICS, leads to further confusion, and, in the long run, prevents the initial responders from developing good decisionmaking skills.

3. **Recommendation:** Onscene personnel should have worn the appropriate ICS position vests.

4. **Recommendation:** The Communications Manager should not have become the primary radio operator for the incident. This would have allowed the Dispatch Manager to concentrate on management duties.

5. **Recommendation:** A PIO should have been appointed earlier in the incident. This may have prevented rumors and inaccurate information being distributed.

---

[22] ibid.

[23] The NIMS guidelines define an EMS Strike Team as five ambulances and one supervisor who have common communications with each other.

www.ingramcontent.com/pod-product-compliance
Lightning Source LLC
Chambersburg PA
CBHW081227170526
45165CB00009B/2984